Modelling for Water Resource Recovery

In Focus – Special Book Series

Modelling for Water Resource Recovery

Editors

Nicolas Derlon, Kris Villez, Heather Stewart and Arifur Rahman

Published by

IWA Publishing
Unit 104-105, Export Building
1 Clove Crescent
London E14 2BA, UK
Telephone: +44 (0)20 7654 5500
Fax: +44 (0)20 7654 5555
Email: publications@iwap.co.uk
Web: www.iwapublishing.com

First published 2024
© 2024 IWA Publishing

British Library Cataloguing in Publication Data
A CIP catalogue record for this book is available from the British Library

ISBN: 9781789064827

Contents

doi: 10.2166/wst.2023.175

Editorial: Water resource recovery modelling 2021 (WRRmod2021 conference)

Our society is transitioning fast into the digital age, spurred by development of cheap and new sensing technology, breakthroughs in computing, and development of efficient algorithms for optimization. This transition is also visible in the field of wastewater treatment and is driving new model developments, especially by exploiting the large data sets available with many utilities. Not surprisingly, WRRmod2021 had featured a strong session on 'data-driven models and digitalization' focused on this hot topic. At the same time, engineering practice calls for more robust models for performance evaluation and optimization of both conventional facilities and innovative processes. As a result, the WRRmod2021 program also exhibited sessions on modelling of new process units (e.g., aerobic granular sludge), modelling of the nitrogen cycle, and integrated/plant-wide modelling.

In this issue on the Water Resource Recovery modelling published in *Water Science & Technology* journal, six relevant studies that highlight recent promising model developments were selected.

Three of the selected papers focused on 'data-driven models and digitalization'. Schneider *et al.* (2022) first present an overview of hybrid modelling methodologies applied to wastewater resource recovery facilities (WRRFs), and discuss challenges and research needs for the design and architecture development of such models, good modelling practice, etc. Torfs *et al.* (2022) discuss the use of digital twins, presented as an innovative and powerful technology that has the potential to harness the power of digitalization. Finally, Li & Vanrolleghem (2022) present their work on the development of influent generators, to produce an input time series required for dynamic simulation of WRRF. Together, these studies show early signs of data maturity at the utility level, where sensor data and models can be used online in a synchronized manner.

Two other selected papers presented new models of innovative process units: the aerobic granular sludge process and the short-cut processes for nitrogen removal (i.e., deammonification and nitrite-shunt). Derlon *et al.* (2022) presented a new model of an aerobic granular sludge system that integrates both microbial and hydraulic selection of granules (based on plug-flow feeding, selective removal, etc.). Kirim *et al.* (2022) provided an overview of the state-of-the-art in modelling of the short-cut processes for nitrogen removal in mainstream wastewater treatment and presented future perspectives for directing research efforts in line with the needs of practice.

Finally, a paper by Brouckaert *et al.* (2022) presented a modelling study for the upgrade of a wastewater treatment plant (WWTP), using an incomplete influent data set. In their study, Brouckaert *et al.* thus demonstrate how the application of a pair of modelling tools, a probabilistic influent fractionator, and a simplified steady-state plant-wide model, can help re-combining the influent fractionation, thus offering the best available information by reconciling available data with the firmly established process knowledge.

Guest Editors
Nicolas Derlon IWA
Eawag — Swiss Federal Institute of Aquatic Science and Technology, Dübendorf 8600, Switzerland

Kris Villez IWA
Oak Ridge National Laboratory, Oak Ridge, TN 37831, USA

REFERENCES

Brouckaert, B., Brouckaert, C., Singh, A., Pillay, K., Flores-Alsina, X. & Ikumi, D. 2022 Using plant data to estimate biodegradable COD fractions – case study kwaMashu WWTP. *Water Sci. Technol.* **86** (9), 2045–2058. https://doi.org/10.2166/wst.2022.314.

Derlon, N., Villodres, M. G., Kovács, R., Brison, A., Layer, M., Takács, I. & Morgenroth, E. 2022 Modelling of aerobic granular sludge reactors: the importance of hydrodynamic regimes, selective sludge removal and gradients. *Water Sci. Technol.* **86** (3), 410–431. https://doi.org/10.2166/wst.2022.220.

Kirim, G., McCullough, K., Bressani-Ribeiro, T., Domingo-Félez, C., Duan, H., Al-Omari, A., De Clippeleir, H., Jimenez, J., Klaus, S., Ladipo-Obasa, M., Mehrani, M.-J., Regmi, P., Torfs, E., Volcke, E. I. P. & Vanrolleghem, P. A. 2022 Mainstream short-cut N removal modelling: current status and perspectives. *Water Sci. Technol.* **85** (9), 2539–2564. https://doi.org/10.2166/wst.2022.131.

Li, F. & Vanrolleghem, P. A. 2022 An influent generator for WRRF design and operation based on a recurrent neural network with multi-objective optimization using a genetic algorithm. *Water Sci. Technol.* **85** (5), 1444–1453. https://doi.org/10.2166/wst.2022.048.

Schneider, M. Y., Quaghebeur, W., Borzooei, S., Froemelt, A., Li, F., Saagi, R., Wade, M. J., Zhu, J.-J. & Torfs, E. 2022 Hybrid modelling of water resource recovery facilities: status and opportunities. *Water Sci. Technol.* **85** (9), 2503–2524. https://doi.org/10.2166/wst.2022.115.

Torfs, E., Nicolaï, N., Daneshgar, S., Copp, J. B., Haimi, H., Ikumi, D., Johnson, B., Plosz, B. B., Snowling, S., Townley, L. R., Valverde-Pérez, B., Vanrolleghem, P. A., Vezzaro, L. & Nopens, I. 2022 The transition of WRRF models to digital twin applications. *Water Sci. Technol.* **85** (10), 2840–2853. https://doi.org/10.2166/wst.2022.107.

doi: 10.2166/wst.2022.314

Using plant data to estimate biodegradable COD fractions – case study kwaMashu WWTP

Barbara Brouckaert [a,*], Christopher Brouckaert[a], Akash Singh[b], Kaverajen Pillay[b], Xavier Flores-Alsina[c] and David Ikumi[d]

[a] WASH R&D Centre, University of KwaZulu-Natal, Durban 4041, South Africa
[b] eThekwini Water and Sanitation, Durban, South Africa
[c] Process and Systems Engineering Centre (PROSYS), Department of Chemical and Biochemical Engineering, Technical University of Denmark, Building, 229, Kgs. Lyngby DK-2800, Denmark
[d] Department of Civil Engineering, University of Cape Town, Cape Town, South Africa
*Corresponding author. E-mail: barbara.brouckaert@gmail.com; brouckae@ukzn.ac.za

BB, 0000-0003-3778-8324

ABSTRACT

A modelling study is under way in preparation for a planned upgrade of the capacity of the kwaMashu WWTP in eThekwini, South Africa, from 50 to 80 ML/d. When the configuration of an existing plant is to be changed, the most critical part of the model calibration is the influent wastewater fractionation. However, the constantly varying characteristics of wastewater make experimental determination of an adequately representative set of components difficult, time-consuming and expensive, which constitutes significant barriers to the adoption of modelling by many municipalities. Compliance and process monitoring generate large sets of influent measurements of chemical oxygen demand (COD), free and saline ammonia (FSA), total suspended solids (TSS), etc., but these are insufficient for modelling purposes. In particular, bio-degradability is not routinely measured. However, since influent fractionation is designed to predict the fate of material in the wastewater treatment process, it should be possible to infer the fractionation from a combination of influent and plant measurements. This case study demonstrates the application of a pair of modelling tools, a probabilistic influent fractionator and a simplified steady-state plant-wide model, to estimate the influent fractionation, together with certain unmeasured or unreliable operational parameters.

Key words: COD fractionation, data reconciliation, parameter identifiability, plant-wide model

HIGHLIGHTS

- Estimation of influent COD fractionation from routine plant measurements.
- Data reconciliation using a simplified steady-state plant-wide model.
- Estimation of unmeasured or unreliable operational parameters.
- Methodology for improving a plant monitoring programme.

GRAPHICAL ABSTRACT

ACRONYMS

AD	Anaerobic digester
ADM1	Anaerobic Digester Model No. 1
AS	Activated sludge
BOD	Biological oxygen demand
COD	Chemical oxygen demand
DAF	Dissolved air flotation
FSA	Free and saline ammonia
IS	Inorganic solids
ISS	Inorganic suspended solids
MLE	Modified Ludzack-Ettinger
MLSS	Mixed liquor suspended solids
OHO	Ordinary heterotrophic organism
Ortho P	Orthophosphate
PST	Primary settling tank
PWM_SA	Plant-wide model – South Africa
RAS	Return activated sludge
SNL	Supernatant liquor from the secondary digester
SRT	Sludge retention time
SSE	Sum of squared errors
SST	Secondary settling tank
TKN	Total Kjeldahl nitrogen
TP	Total phosphorus
TS	Total solids
TSS	Total suspended solids
UCTADM1	University of Cape Town Anaerobic Digester Model No. 1
VFA	Volatile fatty acid
VSS	Volatile suspended solids
WAS	Waste activated sludge
WWTP	Wastewater treatment plant

SYMBOLS

COD_f	Filtered COD
COD_{up}	Unbiodegradable particulate COD
COD_{us}	Unbiodegradable soluble COD
$C_xH_yO_zN_aP_b$	Stoichiometric formula for a generic organic compound
f_{codf}	Soluble (filtered) fraction of influent COD
f_{codup}	Unbiodegradable particulate fraction of influent COD
f_{codus}	Unbiodegradable soluble fraction of influent COD
$f_{cv,i}$	COD to VSS ratio of component i
f_{fsa}	FSA fraction of TKN
f_{iss}	Influent ISS to TSS ratio
f_n	Nitrogen content per gram organic
$f_{ns,codup}$	Fraction of influent unbiodegradable particulates in the settled sewage
$f_{ns,iss}$	Fraction of influent ISS in the settled sewage
f_{nsPST}	Fraction of influent TSS in the settled sewage
f_{OHO}	OHO fraction of influent COD
f_p	Phosphorus content per gram organic
f_{tknf}	Soluble fraction of TKN
f_{tpf}	Soluble fraction of TP
f_{tss}	Influent TSS to COD ratio
f_{vfa}	VFA fraction of influent COD
H	Hessian matrix
N_p	No. of fitting parameters
N_d	Number of measured data points
P	Vector of fitting parameters
R_{AD}	Hydraulic retention time of the anaerobic digesters
S_F	Soluble fermentable organics
S_{NH}	Ammonium

S_{PO_4}	Phosphate
S_U	Soluble unbiodegradable organics
S_{VFA}	Volatile fatty acids
TKN_f	Filtered TKN
TP_f	Filtered TP
V_{AD}	Anaerobic digester volume
V_{AS}	Activated sludge reactor volume
X_{BInf}	Biodegradable particulate influent organics
X_{ISS}	Inorganic suspended solids
X_{OHO}	Ordinary heterotrophic organisms
X_{UInf}	Unbiodegradable particulate influent organics
σ	Covariance matrix

INTRODUCTION

When the configuration of an existing plant is to be changed, the most critical part of the model calibration is the influent wastewater fractionation. Furthermore, due to the complexity of the systems involved, model calibration protocols, for example, Reiger *et al.* (2012), typically involve the sequential calibration of the various subsystems starting with the influent characterization. As a result, errors in the influent characterization are propagated through the other calibration steps (Grau *et al.* 2007).

In general, raw sewage chemical oxygen demand (COD) and total suspended solids (TSS) measurements are available from routine monitoring data. However, treatment models require the fractionation of raw COD and TSS into, at minimum, soluble biodegradable (fermentable organics S_F and volatile fatty acids S_{VFA}) and unbiodegradable (S_U) organic components, particulate biodegradable (X_{BInf}) and unbiodegradable organic components (X_{UInf}) and an inorganic particulate component (X_{ISS}). Figure 1 illustrates the fractionation of the influent COD and TSS in the PWM_SA model (Ikumi *et al.* 2015) used in this study. PWM_SA (Plant-wide model – South Africa) includes a component for ordinary heterotrophic organisms (X_{OHO}) in the influent fractionation; however, in practice, it cannot be distinguished from X_{BInf}, so the two components were lumped together for the purposes of this study.

The constantly varying characteristics of wastewater make experimental determination of an adequately representative set of components, using protocols such as those recommended by the IWA Guidelines (Reiger *et al.* 2012), difficult, time-consuming and expensive, which constitutes significant barriers to the adoption of modelling by many municipalities. Biodegradable organic fractions in raw and settled sewage are typically determined via biological oxygen demand (BOD)

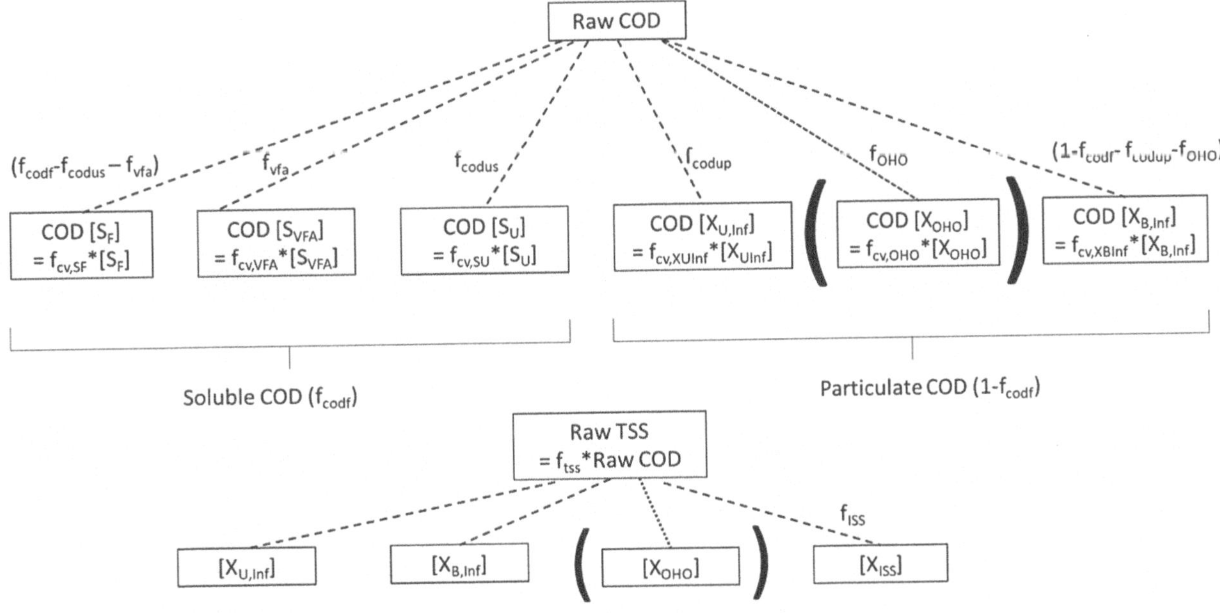

Figure 1 | PWM_SA COD and TSS influent fractionation.

measurements (Hulsbeek *et al.* 2002) or respirometric methods (Vanrolleghem 2002). Both of these methods take days to get results for a single sample, and many municipalities simply do not have the equipment or experienced personnel to undertake these types of characterization studies. Furthermore, translating laboratory results to full scale WWTP plants can be quite challenging due to important differences between the two types of systems (Sin *et al.* 2005).

Compliance and process operation monitoring generates large sets of measurements of COD, TSS, FSA, etc., but these are insufficient for determining the characteristics required by models. Furthermore, they tend to include many errors and inconsistencies, as they are seldom evaluated critically. Nevertheless, a *probabilistic fractionator* tool that we have developed (Brouckaert *et al.* 2016) has proved effective for certain modelling purposes. This combines routine measurements with estimates based on literature and plant experience to determine a probable composition expressed in terms of model components. The probabilistic fractionator, which is included in the PWM_SA model implemented in WEST (Mike powered by DHI 2021), is similar in concept to the over-parametrized influent characterization methodology developed by Grau *et al.* (2007) but includes only the components required for the PWM_SA model as well as a simpler fitting procedure. Figure 2 illustrates the fractionation of influent wastewater measurements into PWM_SA model components for a plant where routine measurements of influent COD, TSS, FSA, orthophosphate, effluent COD and TSS are available. Note that the probabilistic fractionator can be customized to accommodate whatever set of measurements are available and estimate whichever measurements are missing.

With the exception of the soluble unbiodegradable fraction of the influent (f_{codus}), which is typically assumed to equal the soluble COD in the secondary effluent (Ekama & Wentzel 2008), the probabilistic fractionator has no way of distinguishing biodegradable and unbiodegradable fractions. Therefore, the influent COD fractionation parameters (f_{codf}, f_{codp}, f_{vfa}) are usually best guesses based on the available literature.

In the PWM_SA model (Ekama 2009; Ikumi *et al.* 2015), each of the organic components is assigned an elemental composition $C_xH_yO_zN_aP_b$, from which other important component properties can be calculated, including its COD to VSS ratio (f_{cv}), and nitrogen and phosphorus content per gram (f_n and f_p). The biodegradable influent particulate component X_{BInf} is assumed to represent a mixture of proteins, carbohydrates and lipids. These are separate components in some other models, e.g. Anaerobic Digestion Model No1 (ADM1) (Batstone *et al.* 2002); however, since this type of characterization data is very

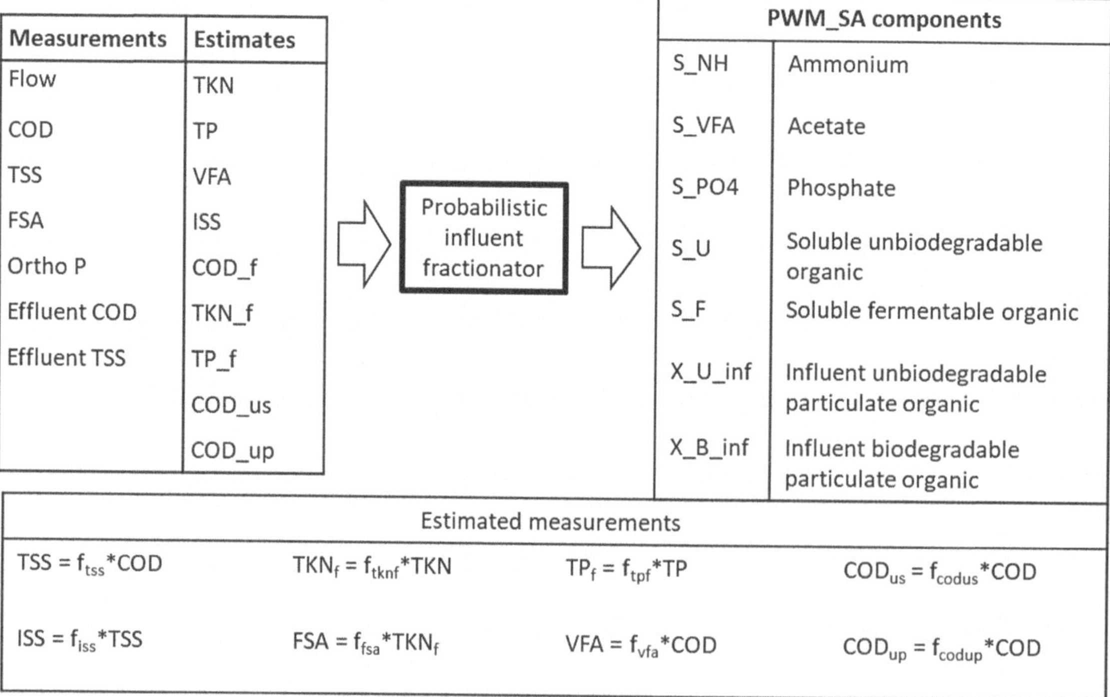

Figure 2 | Schematic representation of the probabilistic influent fractionator showing measurements, estimates and their correlation with the PWM_SA components.

seldom available, they are lumped together in the University of Cape Town Anaerobic Digestion Model No 1 (UCTADM1) (Sötemann *et al.* 2005), which is used to represent anaerobic digestion in PWM_SA. However, the ratio of proteins, carbohydrates and lipids in raw sewage depends on the diet of the population, how much food waste is entering the sewer, and the impact of industrial and commercial wastewater contributions. Therefore, the average composition of the biodegradable particulate fraction may vary significantly between different treatment works. Table 1 shows the elemental compositions for the X_{BInf} model component for various implementations of the plant-wide model compared with ADM1 components for proteins, carbohydrates and lipids (Batstone *et al.* 2002). The listed compositions were all experimentally determined at some point, but it is often not clear which plants the samples they are based on came from and how representative they are of conditions in other plants.

Component compositions are fixed in the probabilistic fractionator, and errors in the composition of X_{BInf} will affect the fitting of the model components to the measured data. In the plant-wide model, the assumed composition of X_{BInf} also affects the predicted release of nutrients in the anaerobic digester, which has implications for nutrient removal.

In this study, we explore the possibility of extending the probabilistic fractionator by including routine process measurements with the routine influent measurements, and coupling the fractionator with a simplified plant-wide steady state model, so as to obtain a more accurate fractionation of the influent. This technique could make wastewater treatment modelling accessible to a wider range of municipalities.

METHODS

kwaMashu wastewater treatment works process description

The kwaMashu Wastewater Treatment Plant (KWWTP) in eThekwini, South Africa, is a conventional wastewater treatment plant with a nominal capacity of 50 ML/d (see schematic representation in Figure 3). It has primary settlers and anaerobic digesters, and is configured for nitrogen but not phosphorus removal. A modelling study is under way in preparation for a planned upgrade to 80 ML/d. The upgrade is planned to improve nutrient removal, and tertiary treatment to recover potable water is also being considered.

The detailed plant layout is shown in Figure 3. After screening and de-gritting, raw sewage is pumped to two primary settling tanks (PSTs). The raw sludge from the PSTs is pumped to two gravity thickeners and the thickened primary sludge is then pumped to the two primary anaerobic digesters. The five-cell secondary digester separates the digested solids from the supernatant liquor (SNL) under gravity. The digested sludge is pumped to the dewatering plant while the SNL is pumped back to the head of works. The primary settler effluent (settled sewage) flows to a Modified Ludzack-Ettinger (MLE) activated sludge process consisting of a four-lane aeration unit followed by eight secondary settlers (SSTs). The secondary effluent is discharged to a series of three maturation ponds and the pond effluent is chlorinated prior to discharge to the Umhlangane River.

The waste activated sludge (WAS) is drawn from the return activated sludge (RAS line). The WAS is pumped to the DAF units for thickening. The DAF sludge is sent to the dewatering plant while the sub-natant flows back to the head of works.

The dewatering plant consists of eight screw presses with polymer dosing. The filtrate from dewatering is returned to the head of works while the dewatered sludge is supposed to be trucked to offsite disposal. However, during the study period, sludge was not being removed from the site on a regular basis due to problems with the disposal contracts and consequently, the operating staff were forced to reduce the sludge wasting rate. As a result, the plant was often being operated at sludge ages of greater than the 12 to 20 days the plant was designed for. Furthermore, since the wasting rate data had been lost when the data storage system was damaged, the actual SRTs during the study period were not known.

Table 1 | Composition of biodegradable influent particulates

| | Wu (2015) | Ikumi (2011) Table 8.2 | WEST PWM_SA Default | Gaszynski (2021) | Flores-Alsina *et al.* (2019) | | | |
					X_{BInf}	Protein	Carbohydrate	Lipid
f_n	0.035	0.03	0.035	0.03–0.05	0.06	0.15	0	0
f_p	0.005	0.01	0.012	0.001–0.004	0.01	0	0	0.03
f_{cv}	1.52	1.47	1.478	1.49–1.52	1.79	1.35	1.07	2.81

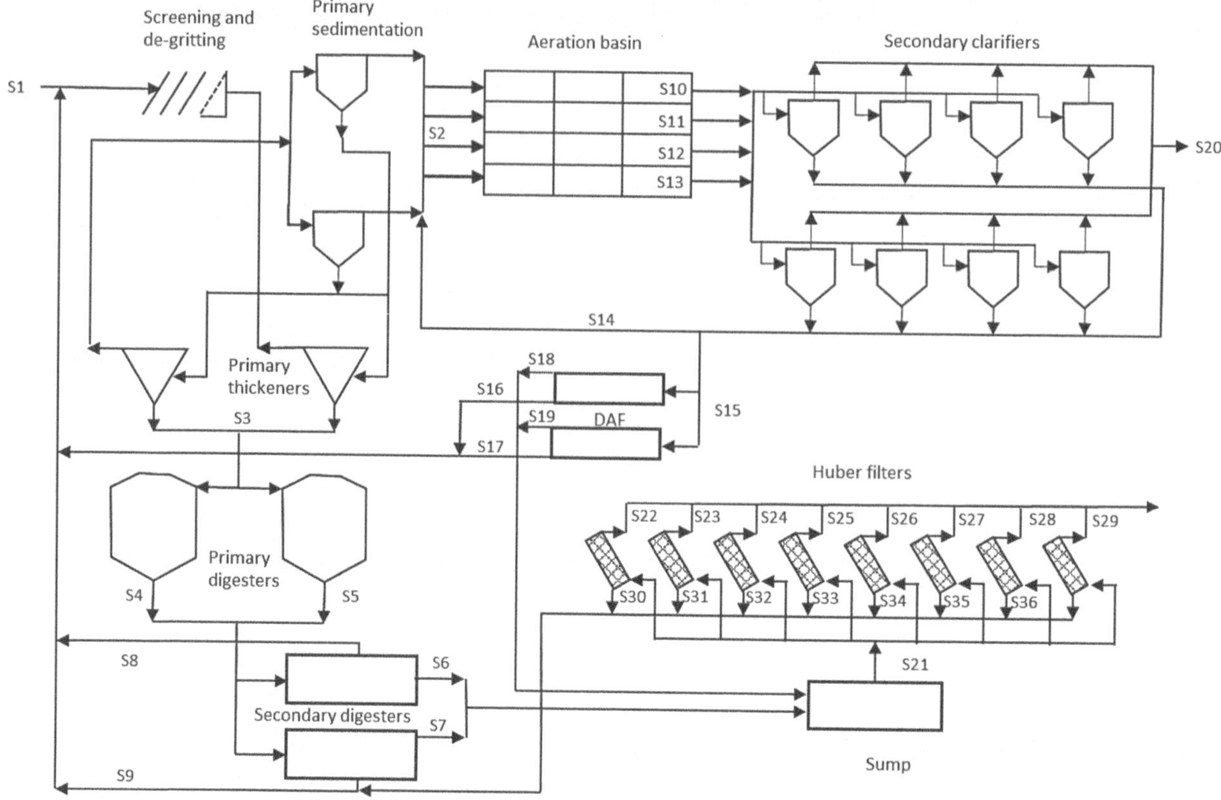

Figure 3 | Detailed *kwaMashu* plant layout with sampling points.

Data collection

Routine monitoring data for 2018 were provided by eThekwini Water and Sanitation (EWS) from their laboratory information system and plant operating records. June and July were a period of little rain and relatively stable plant operation, so these data were selected for modelling. Table 2 shows the plant sampling schedule for the sampling points shown in Figure 3. Influent flowrate was the only measurement made every day. Additional plant operating data recorded by the SCADA system had been lost when the data storage system was damaged by power disruptions.

In addition to the data listed in Table 2, sludge cake production data was available as monthly totals.

Simplified steady state model development

A simplified version of the plant-wide steady state model (Ekama 2009) was set up to track the organic and particulate components through the plant, from the raw sewage to the secondary effluent and sludge filter cake. The purpose of the model was threefold:

Table 2 | Sampling schedule for kwaMashu WWTP

Sampling point	MON	TUE	WED	THU	FRI
Raw sewage S1	pH, TSS, TS, NH$_3$, COD	–	TSS, NH$_3$, COD, PO$_4$	–	pH, TSS, NH$_3$, COD
Settled sewage S2	TSS, NH$_3$, COD	–	TSS, NH$_3$	–	sett solids, susp solids, NH3, COD
Aerated basins S11–S13	pH, TSS, NO$_2$, NO$_3$, NH$_3$	TSS	TSS, NO$_3$, NH$_3$	TSS	pH, TSS, NO$_2$, NO$_3$, NH$_3$
Secondary effluent S20	COD, TSS	COD, TSS	–	COD, TSS	–
Digester feed S3	%TS	%TS	%TS	%TS	%TS
Primary digesters S4, S5	–	%TS, NH$_3$	%TS	%TS	%TS
Filter cake S22–S29	%TS, %IS	%TS, %IS	%TS, %IS	%TS, %IS	%TS, %IS

(1) To estimate the missing internal flows; specifically: the flow to the anaerobic digesters (ADs) and the waste activated sludge (WAS) flow, which determine the hydraulic retention (R_{AD}) time of the ADs and the sludge retention time (SRT) of the activated sludge (AS) plant, respectively.

(2) To estimate the biodegradable organic fractions of the influent based on the overall reduction in solids and COD.

(3) To estimate f_{cv} and f_n of the biodegradable influent particulate component X_{BInf}.

Figure 4 shows the simplified steady state model layout for kwaMashu WWTP.

The influent sub-model for the simplified steady state model includes only COD and particulate components, i.e. the components shown in Figure 1.

In the simplified steady state mode, the PST sub-model lumps together the primary settlers and primary sludge thickeners such that its outputs are the settled sewage and the feed to the anaerobic digesters. The PST sub-model parameters are f_{setsew}, f_{nsPST}, $f_{ns,codup}$ and $f_{ns,iss}$, where $(1 - f_{setsew})$ is the flow to the anaerobic digesters as a fraction of the plant influent flow and determines the anaerobic digester hydraulic retention time (R_{AD}), while f_{nsPST}, $f_{ns,codup}$ and $f_{ns,iss}$ are the settled sewage TSS, X_{UInf} and ISS fluxes as a fraction of their corresponding fluxes in the influent.

The anaerobic digester model is based on Sötemann *et al.* (2005). The simplified anaerobic digester model lumps together the primary and secondary digester models such that the output of the AD sub-model is the thickened digested sludge fed to the dewatering plant. The retention time (R_{AD}) of the anaerobic digester is calculated as

$$R_{AD} = \frac{V_{AD}}{Raw\ flow * (1 - f_{setsew})} \tag{1}$$

where

$V_{AD} =$ digester volume, m^3

Raw flow = raw sewage flow to the primary settlers, m^3/d

The simplified activated sludge (AS) sub-model is based on the steady state model equations for organic material removal presented in Ekama & Wentzel (2008). The simplified model lumps together the activated sludge reactors, secondary settlers and DAF thickening of the waste activated sludge (WAS) such that the outputs of the AS sub-model are the secondary effluent and DAF thickened sludge fed to the dewatering plant, as shown in Figure 4. The sludge age for the activated sludge plant is calculated as

$$SRT = \frac{V_{AS} * MLSS}{(Wasting\ rate + Effluent\ solids\ flux)} \tag{2}$$

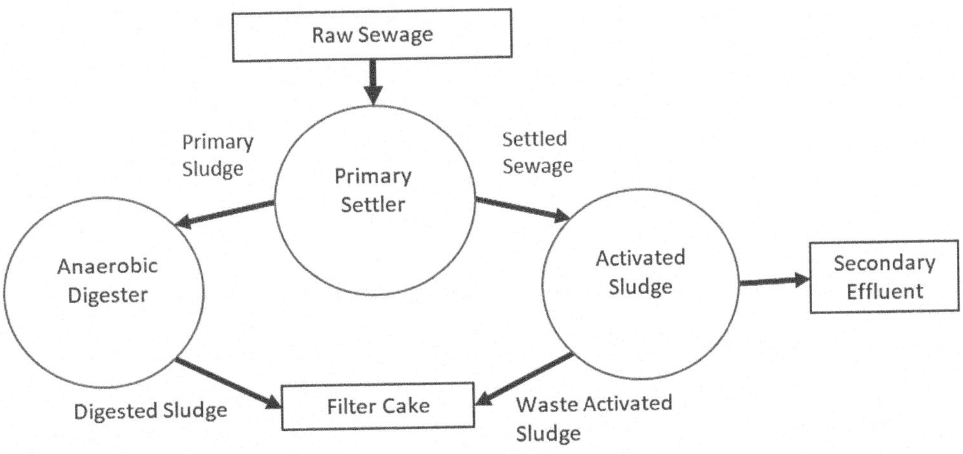

Figure 4 | Simplified plant-wide steady state model.

where

SRT = sludge age, d

V_{AS} = activated sludge reactor volume, m^3

$MLSS$ = mixed liquor suspended solids, kg/m^3

Wasting rate = DAF sludge solids flux fed to sludge dewatering, kg/d

Effluent solids flux = secondary effluent flux, kg/d

Note that is assumed that the impact of recycling the secondary digester supernatant, DAF underflow and dewatering plant filtrate on the flow and solids fluxes have a negligible effect on the overall flow and solids balances. Since the simplified steady state model does not include nutrient removal, the impacts of these internal recycles on the nitrogen and phosphorus fluxes are not considered.

Model implementation, fitting and parameter identifiability

The simplified model and optimization were coded in R (R Project 2021). The standard R non-linear optimization routine *optim* was used for the parameter regression.

The plant measurements used in the fitting procedure and initial set of fitting parameters are listed in Table 3. Outliers in the raw measurements were removed using the interquartile range and fences method.

Some of the parameters listed in Table 3 can be estimated directly from the available raw plant data, using simple mass balances on individual treatment units where necessary, specifically, the fitted raw COD, secondary effluent solids flux, f_{setsew}, f_{tss}, f_{codus} and $f_{ns,PST}$. However, the results obtained this way may be subject to errors and systemic biases, e.g. due to data sparsity, how and when samples are collected and how averages are calculated. Furthermore, uncertainties increase when parameter values are interdependent. For the above mentioned parameters, the raw data was used to calculate the initial estimates and expected ranges. The initial estimates for the fitted raw COD and effluent solids flux were calculated as the average of the available measurements. The fraction of the raw flow going to the activated sludge plant (f_{setsew}) was estimated from a solids balance on the PST using the average raw and settled TSS and thickened primary sludge %TS. The initial estimate of f_{tss} was the ratio of the average measured raw COD and TSS. The initial estimate of the soluble unbiodegradable COD fraction (f_{codus}) was calculated from the average secondary effluent COD and TSS as described by Ekama & Wentzel (2008). The non-settlable influent solids fraction ($f_{ns,PST}$) was estimated as the ratio of the average measured settled sewage and raw TSS.

The upper and lower fitting bounds of the wasting rate corresponded to SRT values of 12–30 days. Initial estimates and ranges for the settled sewage fractionation parameters $f_{ns,codup}$ and $f_{ns,iss}$ were specified based on typical values of 0.2–0.3

Table 3 | Fitted measurements and fitted parameters

Fitted measurements	Fitting parameters
Raw COD flux	*Overall mass balance:*
Raw TSS flux	Secondary effluent solids flux, Fitted raw COD
	Internal flows:
Settled sewage COD flux	Activated sludge wasting rate
Settled sewage TSS flux	Flow split between AS and AD (f_{setsew})
	Influent fractionation:
Secondary effluent COD flux	Influent COD/TSS ratio (f_{tss})
Secondary effluent TSS flux	Unbiodegradable soluble COD fraction (f_{codus})
	Soluble COD fraction (f_{codf})
Reactor MLSS	Unbiodegradable particulate COD fraction (f_{codup})
	Settled sewage fractionation
Digester % TS	Fraction of influent solids in the settled sewage ($f_{ns,PST}$)
Digester FSA	Fraction of influent ISS in the settled sewage ($f_{ns,iss}$)
	Fraction of influent unbiodegradable particulate COD in the settled sewage ($f_{nscodup}$)
Primary sludge % TS	*XBinf composition*
	Biodegradable influent particulate gCOD/g ($f_{cv,XBinf}$)
Dry sludge cake produced (as dry solids)	Biodegradable influent particulate gN/g ($f_{n,XBinf}$)
Sludge cake % IS	

and 0.1–0.2, respectively, reported by Wentzel *et al.* (2006). Initial estimates and bounds for the other fractionator parameters were selected based on values in Henze & Comeau (2008) and typical values for various South African wastewater treatments included in a plant-wide steady state design (PWSSD) tool developed by Wu (2015).

Parameter estimation for the simplified steady state model is illustrated in Figure 5.

A parameter identifiability analysis was carried out using sequential regression of a series of nested models. The identifiability analysis procedure is summarized in Figure 6. The sequence started with regressing the complete set of model parameters. At each subsequent step, the least significant parameter was eliminated, and the regression repeated.

At each iteration of the parameter identifiability analysis, the best fit parameter estimation solution was determined by minimizing the weighted sum of square errors:

$$SSE = \sum_i \sum_j w_i \left(\frac{x_{i,j} - X_i}{X_i} \right)^2 \tag{3}$$

where

wi = weight of measurement i

$x_{i,j}$ = jth observation of measurement i

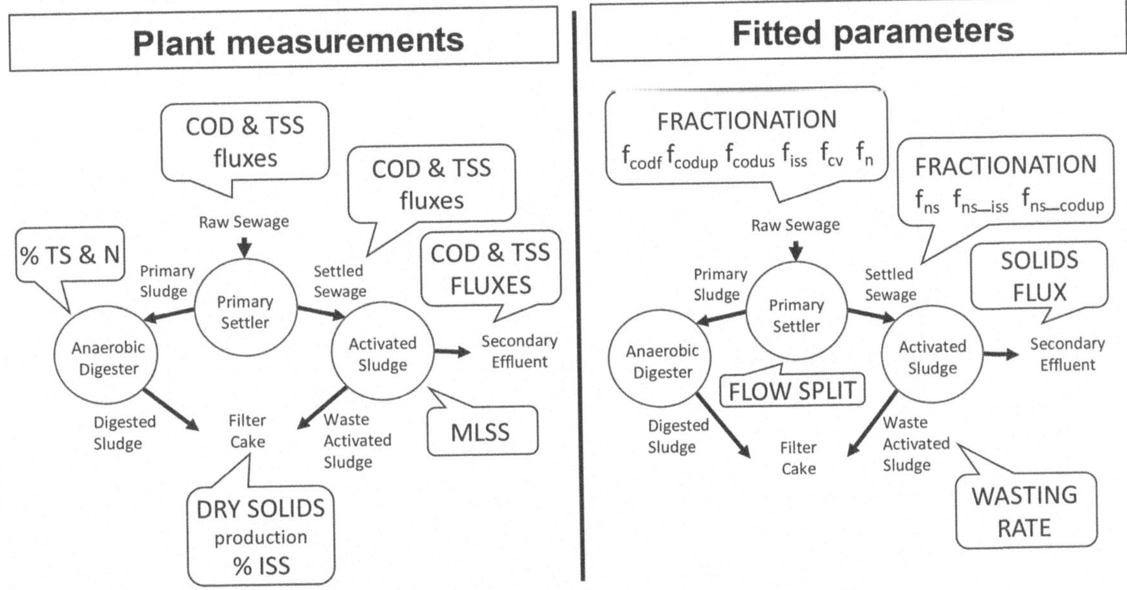

Figure 5 | Parameter estimation for the simplified steady state model.

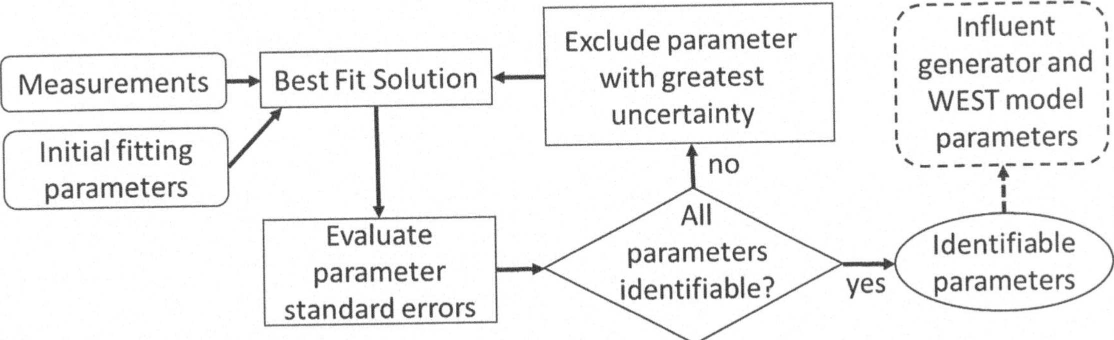

Figure 6 | Parameter identifiability analysis.

X_i = predicted steady state value of measurement x_i

To estimate the significance of the regressed parameters P, the parameter covariance matrix σ was calculated from the inverse of the Hessian matrix H at the optimum point where $H = \partial^2 SSE/\partial P^2$ (Vanrolleghem *et al.* 1995). H is a measure of the local model sensitivity to the parameters at the optimum point, and is an output of the *optim* routine, and $\sigma = SSE/(N_d - N_p - 1) \cdot (H/2)^{-1}$, where N_d is the number of measured data, and N_p is the number of regressed parameters. The significance of parameter i was expressed as its standard error $\sqrt{\sigma_{i,i}}/P_i$: the ratio of its estimated standard deviation to its value. This approximate procedure implies linearizing the model in the vicinity of the optimum point.

If the standard error in any parameter exceeded 0.5 (corresponds to a 95% confidence interval for the parameter that includes the value zero) then the parameter with the greatest standard error was excluded from the fitting, and the regression repeated with the remaining parameters, until the standard errors of all remaining parameters were less than 0.5. These remaining parameters were deemed to be observable from the available data, and their optimized values were then transferred to the probabilistic influent fractionator and the detailed model of kwaMashu WWTW in WEST (Mike powered by DHI 2021) (see Figure 7).

RESULTS AND DISCUSSION

The results of the parameter regression are summarized in Tables 4 and 5. Where a parameter was found to be identifiable in the analysis, the best-fit value is listed. Table 4 also shows the % change from the initial values calculated from the raw plant data.

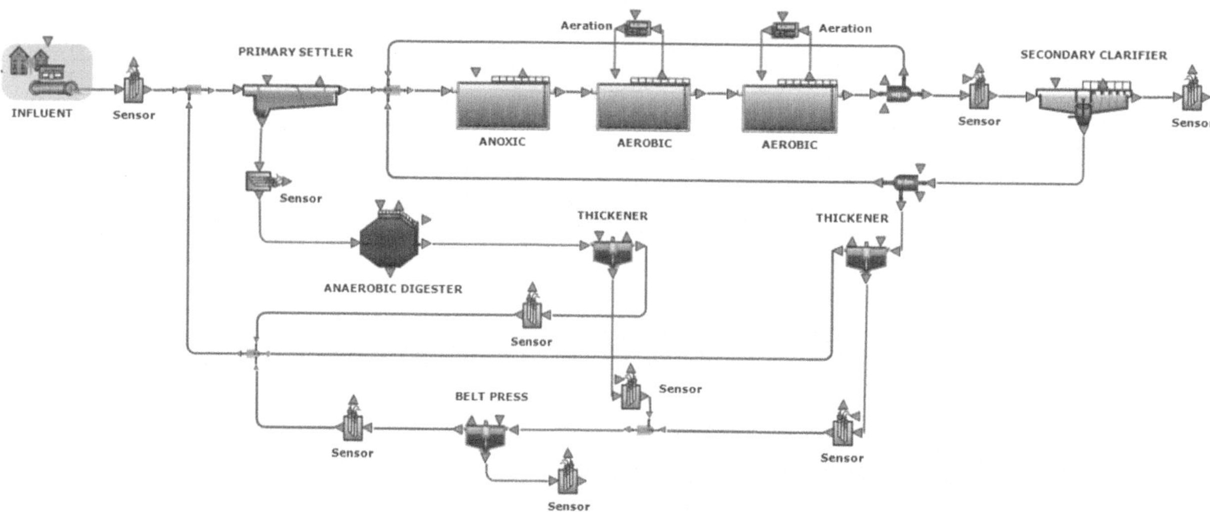

Figure 7 | Detailed WEST model layout of the kwaMashu WWTP.

Table 4 | Parameters which could be estimated directly from raw data

Parameter	Initial estimate	Range	Best estimate	% adjustment	Standard deviation
Secondary effluent solids flux, kg/d	730	464–1,158	738	1%	36
Fitted raw COD, mgCOD/L	830	753–908	850	−2%	43
$(1 - f_{setsew})$	$(1 - 0.993)$	$(1 - 0.97)$–$(1 - 0.995)$	$(1 - 0.994)$	−1%	0.00040
f_{tss}	0.62	0.4–0.65	0.58	−7%	0.034
f_{codus}	0.026	0.01–0.05	0.020	−23%	0.0034
$f_{ns,PST}$	0.267	0.2–0.4	0.315	18%	0.025

Table 5 | Parameters which could NOT be estimated directly from raw data

Parameter	Initial estimate	Range	Best estimate	Standard deviation
Activated sludge wasting rate, kg/d	3,600	1,300–4,500	2,709	283
f_{codf}	0.255	0.1–0.5	–	–
f_{codup}	0.13	0.07–0.2	0.108	0.020
$f_{ns,iss}$	0.03	0–0.3	–	–
$f_{ns,codup}$	0.18	0.1–0.4	–	–
$f_{cv,XBinf}$	1.5	1.47–1.79	1.54	0.11
$f_{n,XBinf}$	0.035	0.03–0.05	0.033	0.0075

It was possible to obtain best fit values for the wasting rate, flow and solids flux split between the anaerobic digesters and AS plant ($(1 - f_{setsew})$ and $f_{ns,PST}$ in Table 4) and the X_{BInf} composition parameters ($f_{cv,XBinf}$ and $f_{n,XBinf}$ in Table 5). The best fit values for $f_{cv,XBinf}$ and $f_{n,XBinf}$ corresponded most closely to biodegradable particulate compositions reported by Gaszynski (2021) and these were used to estimate the X_{BInf} stoichiometry used in the probabilistic fractionator and WEST model. The SRT corresponding to the best fit wasting rate was 20 days, which was at the limit of the design range of 12–20 days.

For the raw sewage fractionation, f_{codus} and f_{codup} (Table 4) were identifiable, but f_{codf} (Table 5) was not. This means that the multi-parameter regression could not estimate a unique ratio of soluble and particulate biodegradable COD that explained the observed data. For the settled sewage fractionation, $f_{ns,iss}$ and $f_{ns,codup}$ (Table 5) were not identifiable. Inorganic solids (%IS) were measured only in the dewatering filter cake at kwaMashu, and the regression could not determine in which ratio the organic and inorganic inert particulates follow the two flow paths available in the simplified model.

Parameters that were excluded from the regression were fixed at their initial estimates, and these fixed values will affect the best-fit values of the identifiable parameters, and the subsequent calibration of the detailed model (Brun *et al.* 2002). However, the results of the analysis also point to a simple and practical solution to the problem: of the three parameters that were not observable from the available data, only $f_{ns,codup}$ requires specialized analytical techniques for its measurement. Any wastewater laboratory should be able to conduct measurements for filtered COD to reduce the uncertainty in the estimation of the parameter f_{codf}. Similarly, the estimation of $f_{ns,iss}$ can be improved by measuring the %IS of the primary, digested or waste sludge or the ISS of the settled sewage. In fact, it is quite common to measure the inorganic or ash content of sludge fed to anaerobic digesters.

The focus of this study was on the practical identifiability of the parameters of interest based on the data, and their structural identifiability was not assessed separately. The unbiodegradable particulate fraction is important for determining the solids concentrations in biological reactors. However, in the steady state model, the mass of inert particulates, in either the anaerobic or activated sludge reactor, is given by the product of the flux of inerts into the reactor and the retention time (R_{AD} or SRT). For example, the mass of inert organic particulates in the activated sludge reactor is given by:

$$M(X_{Uinf}) = F(X_{Uinf}) * SRT \tag{4}$$

where
    $M(X_{Uinf})$ = mass of X_{Uinf} in the reactor
    $F(X_{Uinf})$ = flux of X_{Uinf} fed to the reactor

While the wasting rate and effluent solids flux, which are used in calculating the SRT, are identifiable in the regression, the corresponding flux of X_{Uinf} calculated by the model will be very sensitive to variability in the regressed values of these parameters. In fact, when the value of the wasting rate was manually fixed in the regression, the parameter $f_{ns,codup}$ became identifiable. Since the SRT is a critical operating parameter, the sludge wasting rate ought to be part of routine plant records, but was not available in this case. The results of the identifiability analysis suggest that with reliable sludge wasting rate data and some additional inorganic solids measurements, even the fractionation of the settled sewage at kwaMashu WWTP can be estimated using the simplified steady state model.

Note that, in principle, the multivariate parameter estimation could be carried out in WEST using the full plant-wide model; however, it is computationally intensive, and will take many hours to converge to a solution. The simplified steady state model provides a convenient and much faster way (a few minutes at most) to find the best fit parameters for the COD and solids balances. This then reduces the number of parameters that require tuning in the detailed model, for example, relating to the nitrogen removal model.

Figure 8 summarizes the fit of the detailed WEST model, based on the best fit parameter values obtained from the R code, and Figure 9 the resulting distribution of material between the various plant streams.

As noted above, inorganic solids (IS) were only measured in the belt press filter cake at kwaMashu WWTP. While ISS is usually only a small fraction of the TSS of raw wastewater, it makes up a substantial part of the overall filter cake production (~40% in Figure 8), which is why including ISS in the raw wastewater characterization is important.

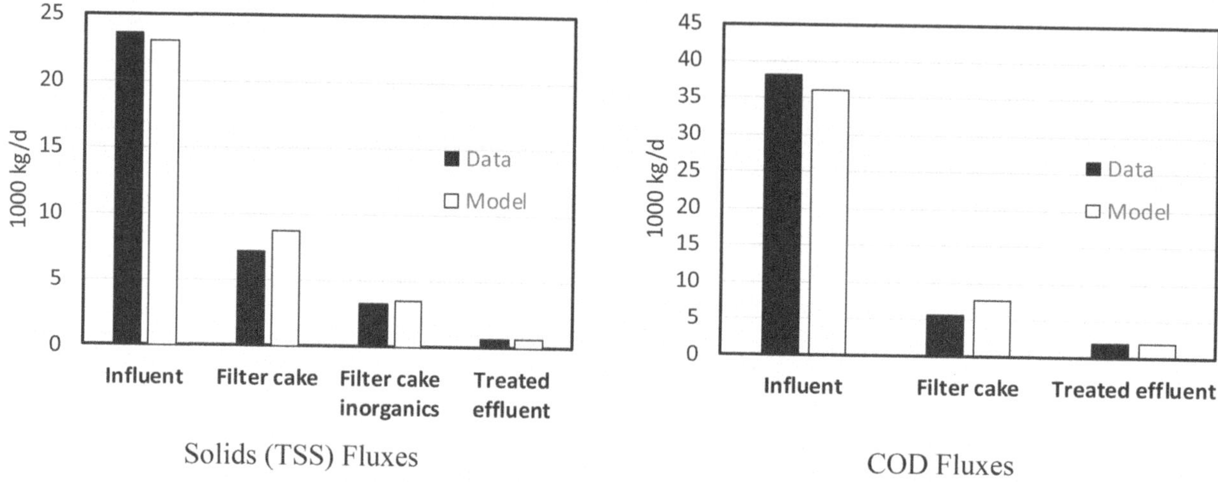

Figure 8 | Agreement between model and measured fluxes of solids and COD in the influent, secondary effluent and filter cake.

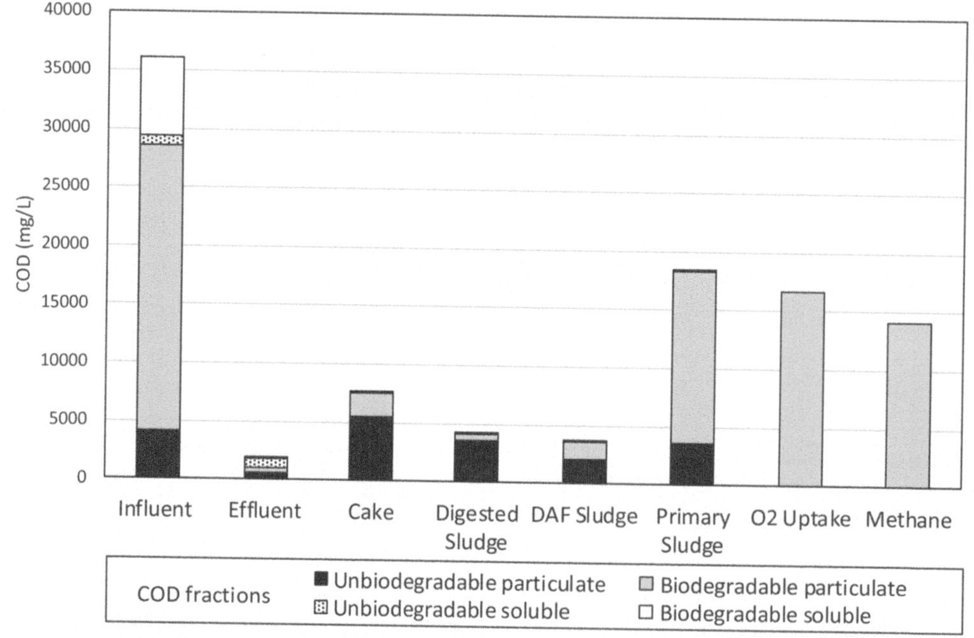

Figure 9 | Model COD balance showing biodegradable and unbiodegradable soluble and particulate fractions.

In Figure 9, the influent COD (S1 in Table 2 and Figure 3), effluent (S20) and sludge cake (S22–29) plus the oxygen uptake and methane production represent the overall plant COD balance, while the primary (S3), digested (S6–7) and DAF thickened waste activated sludge (S18–19) are internal streams. The potential for solids and COD reduction is determined by the biodegradable content of the plant influent, while the extents of the reductions are functions of the sludge age of the AS plant and the hydraulic retention time of the anaerobic digesters, which is why the R tool is able to estimate their values from the observed solids and COD reduction.

As shown in Figure 9, all of the soluble biodegradable COD is consumed while the unbiodegradable soluble fraction mostly exits in the effluent. The bulk of the biodegradable influent particulate is fed to the anaerobic digester in the primary sludge and consumed while the unbiodegradable influent particulate plus some unbiodegradable residue of biomass endogenous respiration ends up in the sludge cake. The biodegradable particulate fraction of the DAF sludge consists of biomass (OHOs) rather than influent organics but makes up only 43% of the COD compared to 68% in the influent.

Particulate COD in the effluent is a small fraction of the total COD balance in Figure 9 but makes up 53% of the secondary effluent COD. This was due to the accumulation of solids in the reactor as a result of the long solids retention time leading to the overloading of the secondary clarifiers.

CONCLUSIONS

We have developed a modelling tool which estimates COD fractionation and other mass balance parameters from plant data using a simplified steady state model of a WWTP. This can greatly facilitate setting up models of treatment plants using routine monitoring data even when some critical operating data, such as the sludge wasting rate, is missing or unreliable. While not all the fractionation parameters investigated were identifiable from the available data in this case study, the results of the significance analysis pointed to practical ways in which data collection could be improved to provide data that is useful for modelling, process optimization and design. Thus, for kwaMashu, it was concluded that all the parameters investigated would likely have been identifiable if the sludge wasting rate records had been available and the routine plant measurements had been augmented by some measurements of filtered raw COD, settled wastewater ISS and/or the IS of the primary, digested or waste activated sludge, as detailed in the discussion. The influent fractionation and calibrated plant model of kwaMashu WWTP will be used by the eThekwini municipality as a benchmark against which the current plant performance is evaluated, and in the design of the proposed upgrade.

ACKNOWLEDGEMENTS

This research was funded by the DANIDA fellowship center through the project 'Evaluation of Resource Recovery Alternatives in South African Water Treatment Systems (ERASE)' (Contract-No: 18-M09-DTU) and by eThekwini Water and Sanitation.

SOFTWARE AVAILABILITY

The probabilistic influent fractionator code and the R implementation of the simplified steady state WWTP model is available on request. To express interest please contact Chris (brouckae@ukzn.ac.za) and Barbara Brouckaert (barbara.brouckaert@gmail.com) at the WASH R&D Centre at the University of KwaZulu Natal (South Africa).

DATA AVAILABILITY STATEMENT

Data cannot be made publicly available; readers should contact the corresponding author for details.

CONFLICT OF INTEREST

The authors declare there is no conflict.

REFERENCES

Batstone, D. J., Keller, J., Angelidaki, I., Kalyuzhnyi, S. V., Pavlostathis, S. G., Rozzi, A., Sanders, W. T. M., Siegrist, H. & Vavilin, V. A. 2002 *Anaerobic Digestion Model No. 1*. IWA Publishing, London, UK.

Brouckaert, C. J., Brouckaert, B. M., Singh, A. & Wu, W. Y. X. 2016 *Wastewater Treatment Modelling for Capacity Estimation and Risk Assessment. WRC Report No. TT 678/16*. Water Research Commission, Pretoria, South Africa.

Brun, R., Kühni, M., Siegrist, H., Gujer, W. & Reichert, P. 2002 Practical identifiability of ASM2d parameters – systematic selection and tuning of parameter subsets. *Water Research* **36**, 4113–4127.

Ekama, G. A. 2009 Using bioprocess stoichiometry to build a plant-wide mass balance based steady-state WWTP model. *Water Research*. **43** (8), 2101–2120. doi:10.1016/j.watres.2009.01.036.

Ekama, G. A., Wentzel, M. C., 2008 Organic matter removal: chapter 4. In: *Biological Wastewater Treatment, Principles, Modelling and Design* (Henze, M., van Loosdrecht, M. C. M., Ekama, G. A. & Brdjanovic, D., eds). IWA Publishing, London, UK.

Flores-Alsina, X., Feldman, H., Monje, V. T., Ramin, P., Kjellberg, K., Jeppsson, U., Batstone, D. J. & Gernaey, K. V. 2019 Evaluation of anaerobic digestion post-treatment options using an integrated model-based approach. *Water Research* **156**, 264–276.

Gaszynski, C. 2021 *Identification of Wastewater Primary Sludge Composition Using Augmented Batch Tests and Mathematical Models.* PhD Thesis, University of Cape Town, Cape Town, South Africa.

Grau, P., Beltrán, S., de Gracia, M. & Ayesa, E. 2007 New mathematical procedure for the automatic estimation of influent characteristics in WWTPs. *Water Science and Technology* **56** (8), 95–106.

Henze, M. & Comeau, Y. 2008 Wastewater characterization: chapter 3. In: *Biological Wastewater Treatment, Principles, Modelling and Design* (Henze, M., van Loosdrecht, M. C. M., Ekama, G. A. & Brdjanovic, D., eds). IWA Publishing, London, UK.

Hulsbeek, J., Kruit, J., Roeleveld, P. & van Loosdrecht, M. 2002 A practical protocol for dynamic modelling of activated sludge systems. *Water Science and Technology* **45** (6), 127–136.

Ikumi, D. S. 2011 *The Execution of a UCT Model of the Proposed Upgrade to the Darvill WWTW. Final Report Prepared for Hatch Goba PTY, South Africa.*

Ikumi, D. S., Harding, T. H., Vogts, M., Lakay, M. T., Mafungwa, H. Z., Brouckaert, C. J. & Ekama, G. A. 2015 *Mass Balances Modelling Over Wastewater Treatment Plants III.* WRC Report No. 1822/1/14, Water Research Commission. ISBN 978-1-4312-0614-8.

Mike powered by DHI 2021 WEST. Available from: https://www.mikepoweredbydhi.com/products/west (accessed 10 January 2022).

R Project 2021 Available from: https://www.r-project.org/ (accessed 10 January 2022).

Reiger, L., Gillot, S., Langergraber, G., Ohtsuki, T., Shaw, A., Takacs, I. & Winkler, S. 2012 *Guidelines for Using Activated Sludge Models. IWA Task Group on Good Modelling Practice.* IWA Publishing, London, UK.

Sin, G., Van Hulle, S., De Pauw, D., van Griensven, A. & Vanrolleghem, P. 2005 A critical comparison of systematic calibration protocols for activated sludge models: a SWOT analysis. *Water Research* **39**, 2459–2474.

Sötemann, S. W., van Rensburg, P., Ristow, N. E., Wentzel, M. E., Loewenthal, R. E. & Ekama, G. A. 2005 Integrated chemical/physical and biological processes modelling part 2 – anaerobic digestion of sewage sludges. *Water SA* **31** (4), 545–568.

Vanrolleghem, P. A. 2002 *Principles of Respirometry in Activated Sludge Wastewater Treatment.* Department of Applied Mathematics, Biometrics and Process Control, University of Ghent, Ghent, Belgium.

Vanrolleghem, P. A., Van Daele, M. & Dochain, D. 1995 Practical identifiability of a bio-kinetic model of activated sludge respiration. *Water Research* **29** (11), 2561–2570.

Wentzel, M., Ekama, G. A. & Sotemann, S. 2006 Mass balance-based plant-wide wastewater treatment models – part 1: biodegradability of wastewater organics under anaerobic conditions. *Water SA* **32** (1), 287–296.

Wu, W. Y. X. 2015 *Development of a Plant-Wide Steady-State Wastewater Treatment Plant Design and Analysis Program.* MSc Thesis, University of Cape Town, Cape Town, South Africa.

First received 10 January 2022; accepted in revised form 20 September 2022. Available online 7 October 2022

doi: 10.2166/wst.2022.220

Modelling of aerobic granular sludge reactors: the importance of hydrodynamic regimes, selective sludge removal and gradients

Nicolas Derlon [a,*], Mercedes Garcia Villodres [a,b], Róbert Kovács [c,d], Antoine Brison [a], Manuel Layer [a], Imre Takács [c] and Eberhard Morgenroth [a,e]

[a] Eawag, Swiss Federal Institute of Aquatic Science and Technology, Dübendorf 8600, Switzerland
[b] WABAG Water Technology Ltd, Bürglistrasse 31, CH-8400 Winterthur, Switzerland
[c] Dynamita, 7 Eoupe, La Redoute, Nyons 26110, France
[d] Nonlineum, 37 Perjes str., Budapest 1165, Hungary
[e] ETH Zürich, Institute of Environmental Engineering, Zürich 8093, Switzerland
*Corresponding author. E-mail: nicolas.derlon@eawag.ch

ND, 0000-0003-4240-9827; AB, 0000-0003-3889-5105; ML, 0000-0002-8426-186X; EM, 0000-0002-1217-269X

ABSTRACT

Hydraulic selection is a key feature of aerobic granular sludge (AGS) systems but existing aerobic granular sludge (AGS) models neglect those mechanisms: gradients over reactor height ($H_{reactor}$), selective removal of slow settling sludge, etc. This study aimed at evaluating to what extent integration of those additional processes into AGS models is *needed, i.e.,* at demonstrating that model predictions (biomass inventory, microbial activities and effluent quality) are affected by such additional model complexity. We therefore developed a new AGS model that includes key features of full-scale AGS systems: fill-draw operation, selective sludge removal, distinct settling models for flocs/granules. We then compared predictions of our model to those of a fully mixed AGS model. Our results demonstrate that hydraulic selection can be predicted with an assembly of four continuous stirred tank reactors in series together with a correction code for plug-flow. Concentration gradients over the reactor height during settling/plug-flow feeding strongly impact the predictions of aerobic granular sludge models in terms of microbial selection, microbial activities and ultimately effluent quality. Hydraulic selection is a key to predict selection of storing microorganisms (phosphorus-accumulating organisms (PAO) and glycogen-accumulating organisms (GAO)) and in turn effluent quality in terms of total phosphorus, and for predicting effluent solid concentration and dynamic during plug-flow feeding.

Key words: aerobic granular sludge, gradients over reactor height, hydraulic selection, modelling, plug-flow feeding

HIGHLIGHTS

- New model for aerobic granular sludge system developed.
- Hydraulic selection of granules (bed stratification, plug flow feeding, etc.) are included.
- Gradients over reactor height predicted with a series of four CSTRs plus plug-flow correction code.
- Concentration gradients over $H_{reactor}$ strongly impact predictions in terms of microbial selection, microbial activities and effluent quality.

GRAPHICAL ABSTRACT

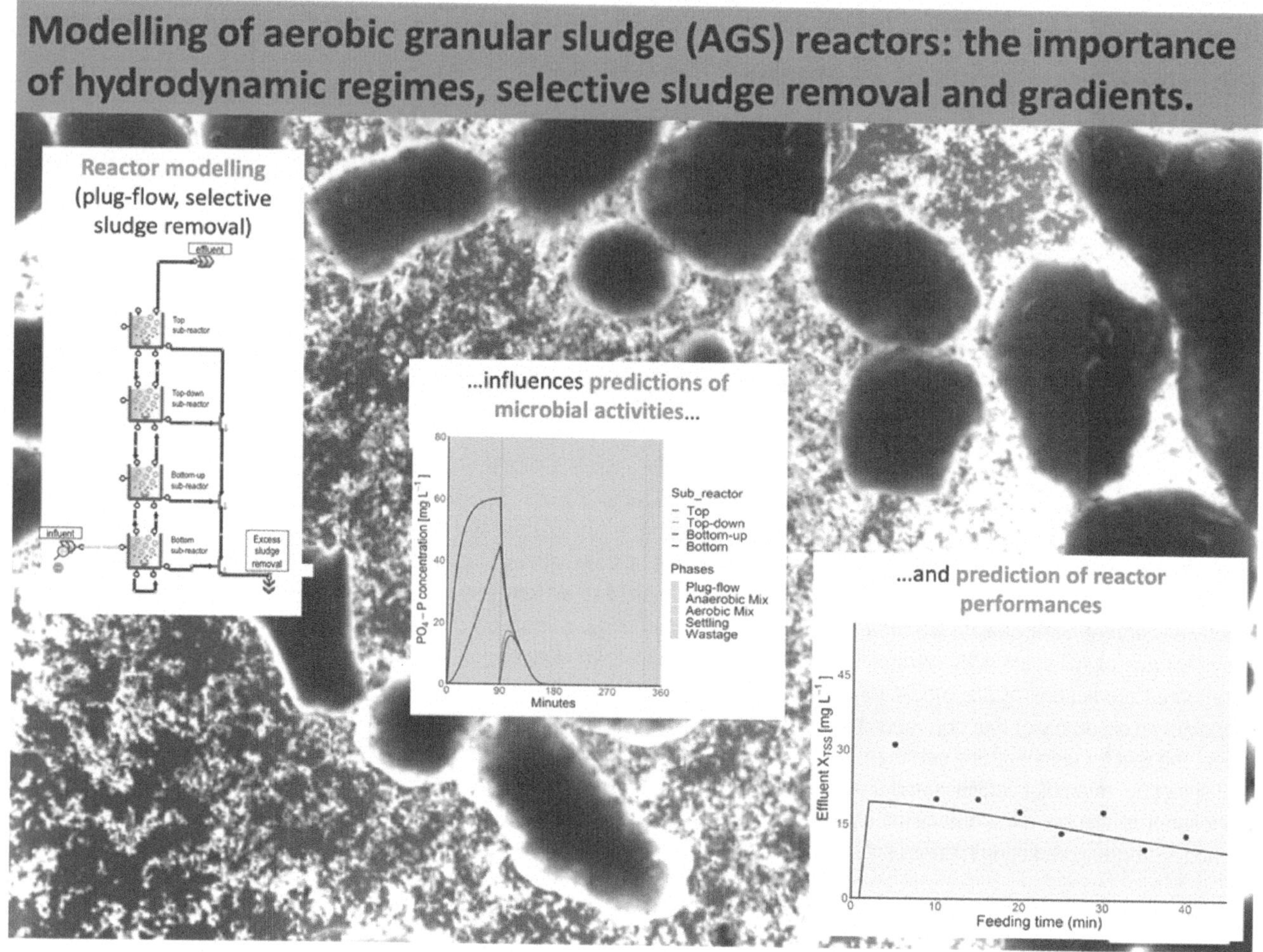

1. INTRODUCTION

Around 80 wastewater treatment plants based on aerobic granular sludge (AGS) and operated as sequencing batch reactors (SBR) are operational or under construction worldwide. However, AGS systems often experience long start-up phases (several months (Pronk *et al.* 2015b; Derlon *et al.* 2016)) or unsuccessful granulation. Also, the design and operation of AGS-SBR is mostly empirical, i.e., based on practical experience. Empirical design criteria are site/situation specific, while the results of dynamic models, when properly used, can be relied upon for generalizable accuracy (Rittmann *et al.* 2018). Dynamic models represent therefore a powerful tool for scientists, engineers, and practitioners. Models can be used by researchers to guide future research efforts and to better understand fundamental mechanisms. Models can also be used by engineers for the planning, design, optimisation, and evaluation of existing or new wastewater treatment plants (WWTP) (Boltz *et al.* 2010). But while the number of AGS systems implemented at full scale is growing rapidly, practicing engineers are still in need of a more appropriate AGS model.

1.1. Experiences from full-scale AGS-based WWTP and implication for model development

Full-scale AGS systems take advantage of both microbial and physical selection mechanisms to form granules. Several key features of AGS systems are still ignored in the structure of existing AGS models, while those mechanisms are of primary importance for granule formation and ultimately for the performances of AGS systems. We refer here to:

(1) the complex composition of AGS, characterized by the coexistence of flocs and granules,

(2) the existence of concentration gradients over the granule depth ($Z_{granule}$),

(3) the existence of concentration gradients over the reactor height ($H_{reactor}$) resulting from the plug-flow feeding and simultaneous fill-draw mode,

(4) the physical selection of granules using the feeding velocity and selective excess sludge removal of slow settling biomass or granules that are too large.

In the following sections we present some lessons learnt from our practical experience with full-scale AGS systems, and provide rationale for their consideration into AGS models.

1.1.1. AGS are hybrid sludge, with both flocs and granules

AGS always contain a small fraction of solids smaller than 200–250 μm[1], referred as 'flocs' in the current study (Pronk *et al.* 2015b; Derlon *et al.* 2016; van Dijk *et al.* 2018; Layer *et al.* 2019). The fraction of flocs in full-scale AGS plants usually varies between 0.1 and 0.2 (based on total suspended solid (TSS) measurements) (Figure 1) and comprises of granule debris, influent particles or biomass growing on organic substrate. AGS systems thus resemble hybrid systems, in which suspended biomass and biofilms coexist. The presence of flocs in AGS influences in turn the effluent quality, e.g., the effluent solids concentration (van Dijk *et al.* 2018). Flocs are also suspected to play an important role in capturing and converting the particulate organic substrates (Pronk *et al.* 2015b; Derlon *et al.* 2016; Campo *et al.* 2020; Layer *et al.* 2020a), thus having a key influence on the proper operation of AGS systems. However, most existing AGS models neglect their presence and in turn their contribution to the system performance (Beun *et al.* 2001; de Kreuk *et al.* 2007; Xavier *et al.* 2007; Ni & Yu 2010; Kagawa *et al.* 2015; Weissbrodt *et al.* 2017). We advocate the structure of AGS models should allow predicting the presence of floccular biomass to best predict the performances of AGS systems.

1.1.2. Gradients over the reactor height ($H_{reactor}$)

Another specific feature of AGS-SBRs is the formation of concentration gradients in both soluble/particulate compounds over the height of the reactor during the non-mixed phases (settling, anaerobic plug-flow feeding). Key for granulation is the selective uptake of substrates by the granules, which results from the stratification of the biomass bed during feeding. A representative bed stratification was characterized at the WWTP of Sarneraatal, Switzerland (Figure 2). Large granules (>2 mm) were found in the first 50 cm over the bottom (i.e., between 650 and 700 cm below surface), where they represented more than 90% of the biomass and reached a very high concentration (>30 gTSS/L). Smaller granules (0.25–2 mm) and flocs were equally observed between 50 and 450 cm above the reactor bottom at concentrations of 2.5–3 gTSS/L. The top of the

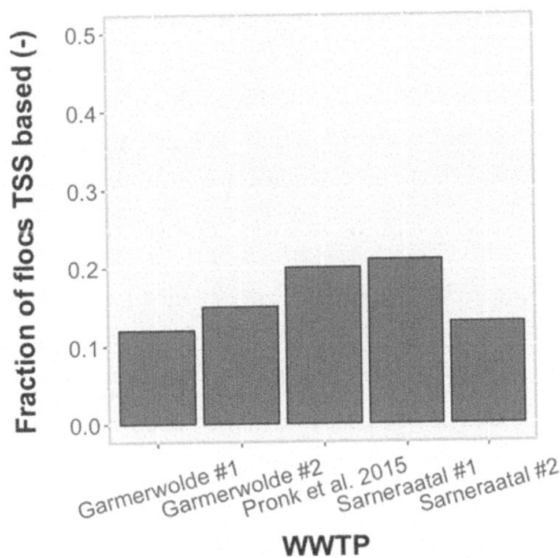

Figure 1 | Fraction of flocs (=solids smaller than 200–250 μm) (based on total suspended solid (TSS)) measured at different full-scale AGS plants.

---

[1] The cut-off between 'flocs' and 'granules' is arbitrary defined and different cut-off values are often reported in literature to make this distinction. However, typical values are usually ranging from 200–250 μm.

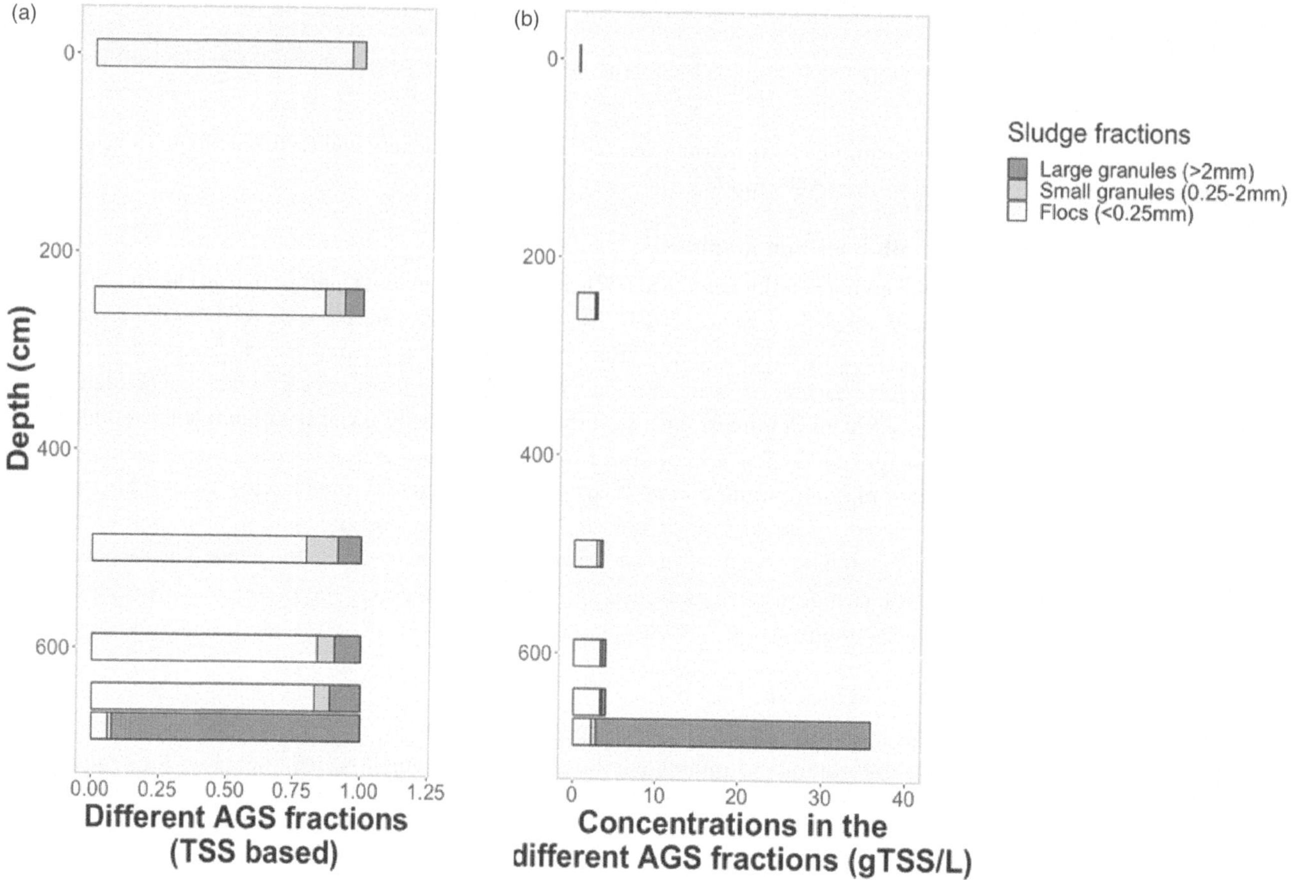

Figure 2 | Stratification of the sludge bed during non-mixed phase (here feeding) measured at the AGS plant of Sarneraatal, Switzerland. (a) different sludge fractions and (b) concentration of the different AGS fractions. 0 cm indicates the surface of the AGS reactor and 700 cm its bottom.

reactor consisted mostly of treated WW, therefore characterized by a very small solids concentration, below legal requirements (<20 mgTSS/L).

Those concentration gradients are essential to provide a competitive advantage to the granules. The distinct settling properties of the flocs and granules govern the stratification of the sludge bed. We advocate that the structure of AGS models (hydraulic reactor model, settling models for flocs and granules) should allow predicting concentration gradients over $H_{reactor}$.

1.1.3. Selective sludge removal

Selective sludge removal is applied at full-scale AGS systems to select granules over flocs (van Dijk *et al.* 2020) and to remove some particulate organic materials originating from the influent, e.g., cellulose (Pronk *et al.* 2015b). Selective sludge removal is usually performed at the top of the sludge bed after sedimentation or feeding, *i.e.*, at a certain depth and after a defined time. Slow settling flocs, small granules or granule debris are removed from the systems through this process (Pronk *et al.* 2015b; van Dijk *et al.* 2020). In some cases, sludge removal is also performed at the reactor bottom to remove large granules. Large granules have a small surface for mass-transfer, which limits their conversion rates. In existing AGS models, sludge removal is often 'forced' to reach an arbitrary fixed solid residence time (SRT) or to maintain granules only into the system. Instead, we suggest sludge removal in AGS reactor model should result from settling properties of sludge and location/time of the sludge withdrawal.

1.2. Limitations of existing aerobic granular sludge models

Many AGS models have been developed and are presented in Table 1. The development of these models clearly advanced our understanding of AGS systems, e.g., spatial distribution and interactions between microbial populations (Beun *et al.* 2001;

Table 1 | Overview of existing AGS models

References	Influent fractionation compatible with municipal WW	Microbial selection based on competition between storing and non-storing microorganisms	Hybrid biomass (flocs and granules)	Granule diameter(s)	Stratification over $Z_{granule}$	Reactor volume (during feeding)	Liquid phase transport	Distinct settling model for flocs/granules	Stratification of the sludge bed	Selective sludge removal
Beun et al. (2001)	–	–	–	Fixed, one class-size	x	Variable	Continuous-stirred tank reactor (CSTR)	–	–	–
Su & Yu (2006)	x	–	x	Fixed, different class-sizes	x		CSTR	–	–	–
de Kreuk et al. (2007)	–	–	–	Fixed, one class-size	x	Variable	CSTR	–	–	–
Xavier et al. (2007)[a]	–	x	–		x	Variable	CSTR	–	–	–
Kagawa et al. (2015)[b]	x	x		Variable, different class-sizes	x	Variable	CSTR	–	–	x[c]
Ni (2013)		x	–	Fixed, different class-size	x	Variable				–
Su et al. (2013)	x	–	x	Variable, different class-sizes	x	Variable	CSTR	–[d]	x[e]	x
Weissbrodt et al. (2017)	–	–	–	Fixed, different class-sizes	x	Constant	Plug-flow and CSTR	–	–	–
Dold et al. (2018)	x	x	x	Variable, one class-size	x	Constant	Plug-flow and CSTR	–[f]	–[f]	x
Eawag AGS model (this study)	x	x	x	Fixed, one class-size	x	Constant	Plug-flow and CSTR	x	x	x

[a]Individual-based model.
[b]Individual-based model.
[c]Selective sludge removal during the 'reactor-scale model' period. Not clear how this selective removal was performed.
[d]Settling model for granules derived from layer model for secondary clarification.
[e]Stratification of the sludge bed predicted, but no plug flow feeding (6 min feeding followed by several hours aeration period).
[f]Only settling of the flocs is predicted. Granules are 'retained' in a bottom sub-reactor during feeding.

de Kreuk *et al.* 2007; Shinya *et al.* 2010), effect of dissolved oxygen (DO) on nitrogen removal (de Kreuk *et al.* 2007; Kagawa *et al.* 2015), etc. Most of these models are therefore able to predict the microbial selection of storing organisms (Lübken *et al.* 2005; Xavier *et al.* 2007; Ni & Yu 2010; Kagawa *et al.* 2015). But while hydraulic selection is also a key factor governing granules formation, only a few models are able to predict the coexistence of flocs/granules (Su & Yu 2006; Su *et al.* 2013; Dold *et al.* 2018), the plug-flow feeding (Weissbrodt *et al.* 2017; Dold *et al.* 2018), the stratification of the sludge bed (Su *et al.* 2013) or the selective removal of slow settling biomass (Su *et al.* 2013; Dold *et al.* 2018). Those aspects are, however, key features of AGS systems, as detailed in section 1.1. More importantly, none of them combine all aspects in link with hydraulic selection of granules, which limit their relevance for engineering practice (design, failure identification, system optimization).

1.3. Objectives

Our work thus aimed at developing an AGS reactor model that combines all aspects related to the microbial and hydraulic selection of granules, therefore correctly representative of the operation and functioning of full-scale AGS plants. A second objective of the work was therefore to demonstrate the importance of concentration gradients over $H_{reactor}$ on the model predictions (microbial population, prediction of effluent quality, etc.), in order to justify such increase in the complexity of AGS model. The goal was not to develop a fully calibrated AGS model validated on a full data set from an AGS-based WWTP. The 'Eawag AGS model' was implemented in Sumo$^{©}$ (Dynamita, Nyons, France) and consisted of a 1-D granule model, a reactor model (allowing for individual settling of granules and flocs and hence sludge bed stratification, plug flow feeding, selective sludge removal based on settling and removal at a specific location) and a bio-kinetic model. Simulations from the Eawag AGS model were compared to the simulations of a conventional fully mixed AGS model.

2. EAWAG AEROBIC GRANULAR SLUDGE MODEL

The Eawag AGS reactor model is an integrated model consisting of (1) a biokinetic model, (2) a granule model and (3) a reactor model. The AGS reactor model was implemented in Sumo$^{®}$ version 20 (Dynamita, Nyons, France) (http://www.dynamita.com). The model is available for download at: https://opendata.eawag.ch/dataset/eawag-ags-model-package. A full description of the model is provided in the package.

2.1. Biokinetic model

The Sumo1 biokinetic model was used in this study (Varga *et al.* 2018). This biokinetic model considers the following microbial populations: ordinary heterotrophic organisms (OHO), nitrifiers (NITO), phosphorus-accumulating organisms (PAO) and glycogen-accumulating organisms (GAO). The following microbial processes are included: (1) growth, (2) decay, (3) hydrolysis, (4) fermentation and (5) storage.

2.2. Granule model

A 1-D Granule model has been developed on the basis of the 'SumoBioFilm' Sumo$^{®}$ model. The 'SumoBioFilm' Sumo$^{®}$ model is a conventional 1-D biofilm model featuring a planar film surface. Our 1-D Granule model considers 2 distinct compartments, the bulk and the granules. The granule compartment is modelled as a sphere sub-divided into n layers (default value: $n = 6$). One main difference between the 1-D SumoBioFilm and the granule model is therefore the area of their layers. The area of the layers is constant over depth for the 'SumoBioFilm' model (planar surface), but variable over depth for the granule model (spherical surface). The diameter of the granules is fixed (input variable). The thickness of the four outer layers is fixed to 25 μm to increase resolution at the granule surface, while inner layers were distributed evenly depending on the chosen granule radius and the total number of layers (Layer *et al.* 2020b, 2022). The thickness of the inner layers is therefore calculated with the following equation:

$$Z_{inner,layer} = \frac{r_{granule} - n_{outer,layer} \cdot Z_{outer,layer}}{n_{total,layer} - n_{outer,layer}}$$

(1)

with $Z_{inner,layer}$ and $Z_{outer,layer}$ the thickness of the inner or outer layers (m), $r_{granule}$ the granule radius (m), and $n_{total,layer}$ and $n_{outer,layer}$ the total number of layer and the number of outer layers, respectively. The number of granules and therefore the overall granule volume is constant (input variable). Different mechanisms are considered in the 1-D granules model, using the same rate equations than the ones used in the 'SumoBioFilm' model: (1) attachment/detachment of particulate compounds, as governed by the concentration gradient between the bulk and top layer of the granules, (2) diffusion of soluble and colloidal compounds from the bulk into the granules and *vice versa*, with inclusion of a mass-transfer boundary layer (3) advection of particulate compounds between the granule layers (referred as transfer and displacement rates in Sumo®).

2.3. Reactor model

2.3.1. Overall structure

A main challenge was to develop a model structure allowing prediction of the stratification over the reactor height. A series of continuous stirred tank reactor (CSTR) can be used to create a pseudo two-dimensional (2-D) model of bulk-liquid hydrodynamics approaching plug flow (Boltz *et al.* 2017). The AGS reactor model thus consists of a series of four child-units assembled into an overall parent AGS reactor model (Figure 3). Each child-unit model is a modified version of Sumo's moving bed biofilm reactor (MBBR) model. This MBBR model represents a biofilm reactor operating at constant volume. The MBBR model was converted into a granules model (see section 2.2.) and additional hydraulic ports and connections were added to connect the four child-units together. Both water and soluble, colloidal and solid compounds (i.e., granules) can flow through those connections and therefore over the reactor height, allowing modelling of settling, granules wash-out with effluent, mixing conditions during mixed aerobic phase, etc. The reactor model additionally allows a variable volume, which is required when selective removal of sludge is applied. Moreover, the reactor model is equipped with an influent port (at the bottom) and an effluent port (at the top), thus allowing for operation in fill and draw mode.

2.3.2. Plug flow optimization

As our reactor model consists of a series of four CSTRs only, it only allows approaching plug-flow conditions. A correction code, active only during feeding, was therefore implemented to mimic an ideal plug-flow. The correction code helps

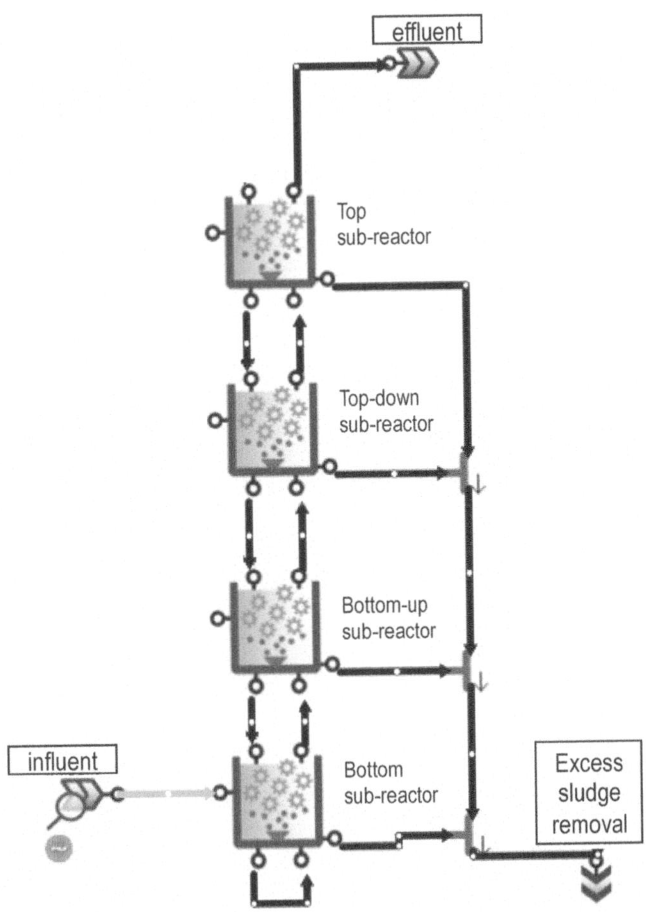

Figure 3 | Conceptual representation of the development of the AGS reactor model. Both water and soluble, colloidal and solid compounds (i.e., granules) can flow through the different connections and therefore over the reactor height.

'retaining' soluble compounds fed into a child-unit until the bulk volume has been fully exchanged. The bulk volume is calculated as the difference between the volume of a child-unit and the volume of granules.

2.3.3. Modelling sludge bed stratification

Stratification of the sludge bed is governed by (1) the settling properties of the granules and flocs, (2) the volume of the sub-reactors and (3) the feeding velocity ($v_{feeding}$). Settling of flocs into the pore space of the bottom child-unit is not possible. Also, the settling of granules (downwards) induces an upwards displacement of bulk (including flocs) towards the upper child-unit. Within each child-unit, the feeding velocity $v_{feeding}$ is calculated based on the available volume of the bulk compartment, i.e., the volume of interstitial voids in a child-unit where granules settled ($V_{reactor} - V_{granules}$) (Equation (2)).

$$v_{feeding} = \frac{Q_{inf}}{A_{reactor} \cdot \dfrac{V_{child-unit} - V_{granules}}{V_{child-unit}}} \tag{2}$$

where $v_{feeding}$ is the feeding velocity (m d^{-1}), Q_{inf} is the influent flow (m^3 d^{-1}), $A_{reactor}$ the reactor surface area (m^2), $V_{child-unit}$ the volume of the child-unit (m^3) and $V_{granules}$ the volume of granules contained in the child-unit (m^3).

An additional important aspect of the model is the volume of each CSTR that is adjusted by modifying the height of the sub-reactor, based on full-scale measurements and in order to reach an ideal stratification of the sludge bed: granules accumulated in the bottom sub-reactor only, for a voidage coefficient of 0.25 ($H_{bottom} = 0.57$ m), flocs in the bottom-up sub-reactor ($H_{bottom-up} = 3.64$ m), and the supernatant above in the top-down and top sub-reactors ($H_{top-down}$ and H_{top} of 2.1 and 0.7 m, respectively). Such adjustment of the sub-unit height/volume is directly derived from knowledge gained from full-scale plants and the cumulative height of the four sub-reactors, thus corresponds to the height of mixed liquor at the WWTP (here 7 m) (Figure 2).

2.3.4. Modelling sedimentation

Distinct models are implemented to predict independently the sedimentation of granules and flocs. The granule settling velocity ($v_{sett,gran}$, m/s) is calculated according to a discrete particle settling model, assuming spherical particles and following the Newton equations (MWH 2012). As can be seen from Equations (3)–(6), there is a circular dependency between the granule settling velocity and the Reynolds number. When implementing the model in Sumo, the tool recognizes these loops and solves them without manual interaction. The calculation of the granule settling velocity depends on the Reynolds number (Re).

$$Re = \frac{\rho_{H2O} \cdot 2 \cdot z_F \cdot v_{sett,gran}}{\eta_{H2O}} \tag{3}$$

If $Re < 2$ (laminar conditions) then:

$$v_{sett,gran} = \left(\frac{g \cdot (\rho_G - \rho_{H_2O}) \cdot (2 \cdot z_F)^2}{18 \cdot \eta_{H_2O}} \right) \tag{4}$$

If $2 < Re < 500$ (transient conditions) then:

$$v_{sett,gran} = \left(\frac{g \cdot (\rho_G - \rho_{H_2O}) \cdot (2 \cdot z_F)^{1.6}}{13.9 \cdot \rho_{H_2O}^{0.4} \cdot \eta_{H_2O}^{0.6}} \right)^{\frac{1}{1.4}} \tag{5}$$

If $Re > 500$ (turbulent conditions) then:

$$v_{sett,gran} = \left(\frac{4 \cdot g \cdot (\rho_G - \rho_{H_2O}) \cdot (2 \cdot z_F)}{3 \cdot c_D \cdot \eta_{H_2O}} \right)^{\frac{1}{2}} \tag{6}$$

where ρ_{H2O} and ρ_G are the water and granule density, respectively, z_F is the granule radius, η_{H2O} is the dynamic viscosity of water, g is the gravitational acceleration constant (9.81 m s^{-2}) and c_D the coefficient of drag (0.44 assuming smooth and rigid spheres). One may acknowledge that aerobic granules are not perfectly smooth and rigid spheres, and that the value of their drag coefficient might be larger than 0.44. However, the settling of aerobic granules mostly occurs under transient hydrodynamic conditions, for which the drag coefficient does not influence the settling velocity predicted by our model (Equation (5)). If the focus is on the prediction of the settling velocity under turbulent conditions, one would have to establish an empirical correlation between the Re number and the C_D coefficient of their granules (van Dijk *et al.* 2018).

The settling velocity of the flocs ($v_{sett,flocs}$, m/s) is calculated according to the Vesilind equation (stock Sumo settling model), which integrates floccular, hindered and compressed settling (Takács *et al.* 1991):

$$v_{sett,flocs} = \max(0;\, min(v_{max};\, compr_{corr} \cdot v_{bnd} \cdot (e^{(-r_{hin} \cdot X_{TSS0})} - e^{(-r_{floc} \cdot X_{TSS0})}))) \tag{7}$$

where v_{max} and v_{bnd} are the maximum Vesilind and boundary settling velocity, respectively, r_{hin} and r_{floc} are the coefficients for hindered and floccular settling, respectively, and $X_{TSS,0}$ the effective X_{TSS} concentration in the bulk phase ($\max(X_{TSS,bulk} - X_{TSS,nonsettleable};0)$. $X_{TSS,nonsettleable}$ is an input variable. The compression term is calculated according to:

$$compr_{corr} = \min(1;\, e^{(r_{compr} \cdot (-X_{TSS,bulk} + compr_{on}))}) \tag{8}$$

where r_{compr} is the coefficient for compression, $X_{TSS,bulk}$ the bulk X_{TSS} concentration and $compr_{on}$ the boundary compression concentration. One may however bear in mind that 'flocs' from AGS systems, i.e., all solids smaller than 250 μm, can contain a large fraction of dense debris resulting from granules breakage. The settling velocity of flocs from AGS systems might therefore exceeds the one of flocs from conventional activated sludge system.

2.3.5. Modelling of mixed conditions

During the mix phases of the SBR cycle (e.g., aerated phase), a very high Q_{mix} flow is applied between the different child-units to artificially create mixing conditions between the four child-units.

2.3.6. Excess sludge withdrawal

Each child-unit is equipped with a port for excess sludge withdrawal (Figure 3). Excess sludge withdrawal can occur through one or several ports. In the default Eawag AGS model, only flocs are withdrawn through excess sludge removal, but the removal of granules is in theory also possible if the model is used with a variable granules number/volume. Sludge wastage can be performed either selectively or based on a target SRT chosen by the user. Selective sludge withdrawal is governed by a combination of variables: settling velocities of the flocs and granules, duration of the sedimentation phase and depth of the port(s) selected for sludge extraction. In this case, a certain volume of bulk liquid is removed from a given sub-reactor, inducing a downward displacement of bulk liquid from the above sub-reactors (flow rates between sub-reactors computed according to the hydraulic balance). When excess sludge is withdrawn based on target SRT, the amount of wasted sludge is calculated according to the target SRT value, to the amount of solids lost with the effluent, and to the total amount of solids in the system. SRT calculation considers both the mass of flocs and of granules in the systems. However, only flocs are wasted (for both excess sludge removal modes). When only one port is used for excess sludge removal, the amount of sludge removed is limited by the amount of TSS in the bulk compartment of the corresponding child-unit. Sludge withdrawal can also be done via the removal of a certain bulk volume. In this case, the SRT is calculated according to the amount of solids withdrawn via excess sludge removal and via effluent.

2.3.7. Default scenario and parameters

Default parameters of the default scenario are listed in Table 2. Simulations were run for 150 d and data of one additional SBR cycle (0.25 d) was then extracted and analysed using R-Studio (Version 3.6.3, 2020).

2.4. Fully-mixed AGS model (reference model)

The fully-mixed AGS model consists of the same building blocks as the Eawag AGS model (a biofilm, biokinetic and reactor model). The granule/biofilm and the biokinetic model is similar to the one used in the Eawag AGS model. The reactor model consists, for the fully AGS model, of a single CSTR which is fully-mixed during all SBR phases and therefore operated in variable volume mode. No settling model is included as the sludge bed stratification is not predicted. Effluent total solids are an

Table 2 | Default SBR cycle parameters, biofilm model parameters and influent WW composition

SBR cycle parameters		Reference
Total cycle duration	6 h	Layer *et al.* (2020b), Layer *et al.* (2022)
Anaerobic feeding phase	Plug-flow, 1.5 h, up-flow velocity of WW during feeding $v_{ww} = 1.67$ m h^{-1}, settling granules/flocs ON	
Aerobic phase	Fully mixed, 4 h, constant dissolved oxygen (DO) concentration = 2.0 mg L^{-1}, settling processes are inactive during the aerobic phase	
Settling phase	Settling processes are active, 0.5 h	
Wasting phase	0.1 h, as part of the settling phase and selectively from second sub-reactor starting from the top of the reactor, to achieve a SRT$_{target}$ = 20 d, settling processes are active during the wasting phase	
Volume-exchange-ratio (VER)	50%	
Hydraulic retention time (HRT)	12 h	
Biofilm model parameters		
Biofilm layers (n)	6	Layer *et al.* (2020b), Layer *et al.* (2022)
Granule radius (z_F)	750 μm	
Thickness of 4 outer granule layers (z_D)	0.25 μm	
Maximum granule X_{TSS} in the system ($X_{TSS,gran,max}$)	6.12 kgTSS m$^{-3}_{reactor}$	Layer *et al.* (2020b), Layer *et al.* (2022)
Maximum X_{TSS} within the biofilm ($X_{TSS,max}$)	102 kgTSS m$^{-3}_{compartment}$	SumoBioFilm model
Granule volume fraction ($X_{TSS,gran,max}/X_{TSS,max}$)	0.06 m$^3_{compartment}$ m$^{-3}_{reactor}$	
Influent composition	Value [unit]	SumoBiofilmModel
Total COD	420 mg L^{-1}	
Total Kjeldahl nitrogen (TKN)	34.4 mg L^{-1}	
Total phosphorus (TP)	4.3 mg L^{-1}	
Filtered COD fraction (incl. colloids, VFA)	40.5%	
Filtered flocculated COD fraction (incl. VFA)	20.2%	
VFA fraction of filtered COD	11.8%	
Unbiodegradable filtered COD fraction	11.8%	
Influent particulate inert COD fraction	14.0%	
Influent heterotrophic fraction of COD	5.0%	
Influent endogenous products fraction of OHOs	20.0%	
Unbiodegradable fraction of influent colloids	20.0%	
Ammonia fraction of TKN	69.8%	
Phosphate fraction of TP	58.1%	
N fraction of filtered biodegradable COD	4.0%	

(Continued.)

Table 2 | Continued

SBR cycle parameters		Reference
N fraction of unbiodegradable COD	1.0%	
P fraction of filtered biodegradable COD	1.0%	
P fraction of unbiodegradable COD	0.1%	

input parameter and are also not predicted. All details including a list of all parameters and default values of the fully-mixed AGS model can be found in Layer *et al.* (2020b, 2022).

2.5. Modelling scenarios

Different simulations were performed to highlight the importance of modelling gradients over $H_{reactor}$ on the prediction of the functioning and performances of AGS reactors (Table 3). Simulations were performed to assess (1) the modelling of the plug-flow feeding (scenario #1), (2) to evaluate the influence of concentration gradients over $H_{reactor}$ on the microbial selection and activities (scenario #2), and (3) to highlight the functionality of the Eawag AGS model to predict effluent quality or optimize operation of AGS-SBR (scenario #3).

3. RESULTS

3.1. Modelling plug-flow conditions (scenario #1)

A main attribute of the Eawag AGS model is to predict plug-flow hydrodynamic conditions during feeding (Figure 4). Plug-flow feeding predicted by our model was compared to predictions by a series of 10 CSTR and 100 CSTR (based on similar hydraulic retention time (HRT) of 1 d), using a non-reactive tracer (Figure 4). In the case of a series of 10 CSTR, breakthrough of tracer starts after 0.4 d only and 1.8 d are required to reach an equal substrate concentration in both the effluent and influent. A better plug-flow behaviour is predicted with a series of 100 CSTR: breakthrough starts after 0.8 d while the effluent trace concentration equals the influent concentration after 1.2 only. However, a perfect plug-flow feeding is predicted only by the Eawag AGS model (four CSTRs in series plus correction code). Breakthrough and equal tracer concentrations in influent and effluent are both observed at 1 d, corresponding to the HRT of the reactor.

Table 3 | Overview of the different simulation scenarios

Scenario	Main focus	Objective(s)	Models/conditions
Scenario #1	Modelling ideal plug-flow conditions during feeding	To evaluate the influence of combining CSTRs in series to model plug-flow conditions	10 CSTRs in series 100 CSTRs in series 4 CSTRs in series + correction code (Eawag AGS model) Non-reactive tracer.
Scenario #2	Impact of modelling concentration gradients over $H_{reactor}$	To demonstrate the importance of concentration gradient over $H_{reactor}$ for the prediction of population distribution, microbial activities and effluent quality.	Fully mixed AGS model: no gradient over $H_{reactor}$. Hydraulic selection is inactive. Eawag AGS model: gradients over $H_{reactor}$ are modelled. Plug-flow regime during settling and feeding (anaerobic) phases. Mixed conditions during the aerobic phase.
Scenario #3	Optimization of total nitrogen removal via optimized aeration	To highlight how the Eawag AGS model can be used to optimizing the operation and performances of AGS systems	Constant setpoint for aeration at 2 mgO_2/L Variable setpoint for aeration (first 2 mgO_2/L and then 0.5 mgO_2/L)

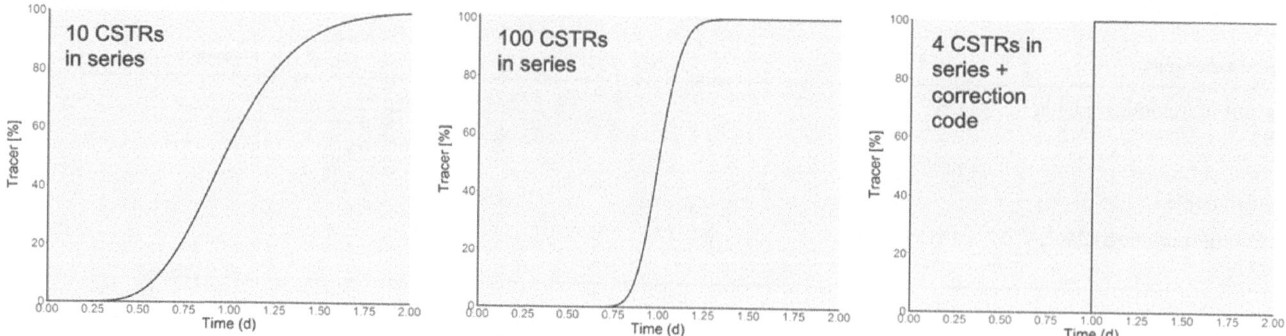

Figure 4 | Modelling of the plug-flow feeding with different model structures: 10 CSTRs in series, 100 CSTRs in series and 4 CSTRs in series plus correction code (as implemented in the Eawag AGS model).

3.2. Modelling of sludge bed stratification

Prediction of the distribution of flocs and granules in the four sub-reactors during the different phases of a default SBR cycle is shown in Figure 5. During the *settling and feeding phases (pink/blue phases)*, granules quickly settle towards the reactor bottom and accumulate exclusively in the bottom sub-reactor (black line), as the granule settling velocity always exceeds the feeding velocity in this default simulation. No granules are found in the upper sub-reactors. Flocs also settle during those *settling and feeding phases*. However, flocs mostly accumulate into the bottom-up sub-reactor, *i.e.*, on the top of the granules, and to a minor extent into the upper sub-reactor (top-down). Floc accumulation in the bottom sub-reactor is limited by the sedimentation of the granules, and the resulting upwards displacement of bulk volume. Floc concentration in the top sub-reactor mirrors the concentration of solids found in the effluent during feeding and is therefore very low. Overall, the sludge bed stratification predicted by the Eawag AGS model matches quite well the ones characterized at full-scale AGS-based plants, as the one of Sarneraatal WWTP shown in Figure 2.

During the mixed phases (green phase), the AGS reactor is then operated as a closed system (influent, effluent and wastage Q are nil) while a certain Q_{mix} flow is applied among the four sub-reactors to operate the AGS reactor consisting of four

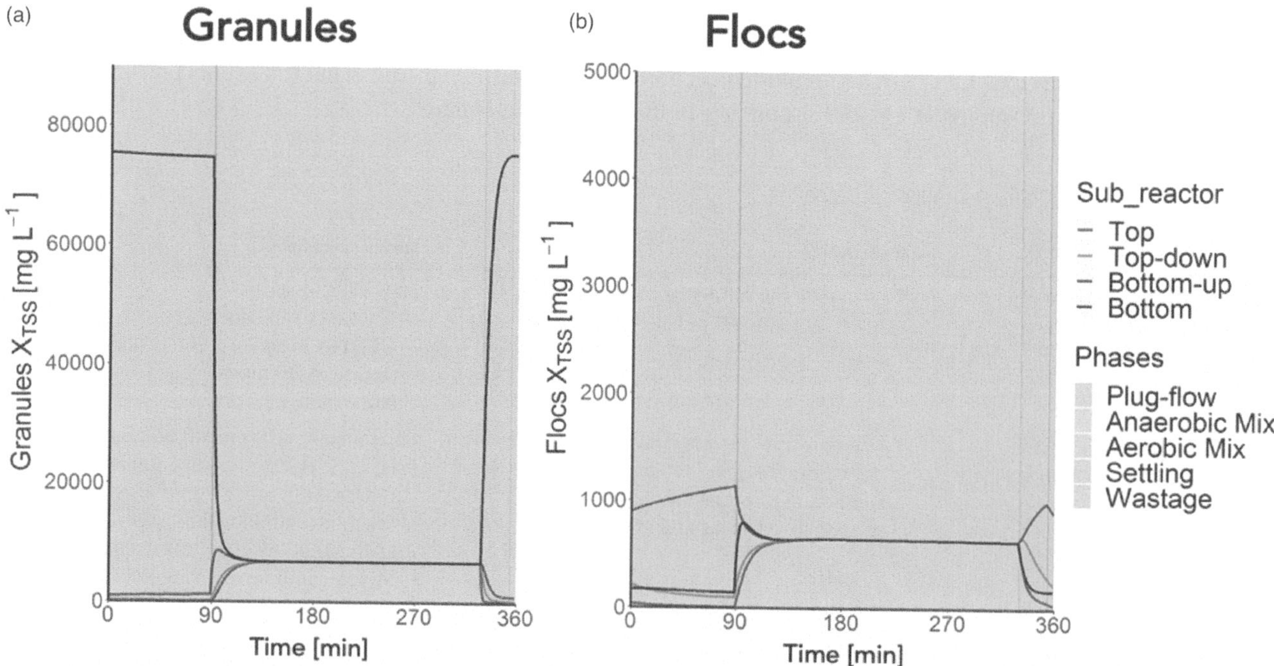

Figure 5 | Distribution of the (a) granules and (b) flocs over the different sub-reactors and different phases of the SBR.

CSTRs as a single CSTR. The flocs and granules concentrations thus equalize in each sub-reactor. During the *wastage phase (purple phase)*, excess sludge withdrawal occurs in the top-down sub-reactor and only flocs are withdrawn.

3.3. How do concentration gradients over $H_{reactor}$ influence microbial activities and population distributions (scenario #2)?

3.3.1. Microbial activities

The influence of modelling stratification over $H_{reactor}$ was further investigated in terms of nutrient profiles (Figure 6) as a result of the different microbial activities (Figure 7).

Predictions by a conventional fully-mixed AGS model: during the mixed non-aerated feeding, nitrate concentration first decreases as a result of dilution and denitrification. Denitrification is taking place in the entire reactor volume due to mixed conditions. Once denitrification is completed and anaerobic conditions are established (t = 35 min.), the release of ortho-phosphates starts due to the activity of PAO and takes place until the end of the feeding phase (90 min). At t = 90 min, ortho-phosphate concentration in the bulk reaches a value of around 10 mgP/L. During the mixed aerobic phase, both nitrification by nitrifying organisms (NITO) and phosphorus uptake by PAO occur in parallel. Complete ammonium removal is achieved after 196 min (based on a legal requirement of 2 mgN/L), while full phosphorus uptake is predicted only after 208 min (legal requirement of 0.8 mgP/L).

Prediction by the Eawag AGS model (SCENARIO #2): the Eawag AGS model allows predicting the microbial activities over $H_{reactor}$ during feeding. An important PAO activity is predicted by the model, as shown by the significant increase in the ortho-phosphate concentration in the granule and floc bed (bottom and bottom-up sub-reactors, respectively). A complete release of ortho-phosphate by the granules accumulated in the bottom sub-reactor is also predicted after 50 min (Figure 6), as a result of the very rapid establishment of anaerobic conditions due to the quick denitrification taking place in the bottom sub-reactor (Figure 6). At the end of the feeding phase (t = 90 min), ortho-phosphate concentrations reach 60 and 45 mgP/L in the granule and floc bed, i.e., in the bottom and bottom-up sub-reactors, respectively (Figure 6, Eawag AGS model – ortho-phosphate profiles). This corresponds to a concentration of 20 mgP/L after a few minutes of mixing during the aerobic phase, as opposed to 10 mgP/L in the bulk predicted by the fully mixed AGS model. During the mixed aerobic phase, the uptake of ortho-phosphate by PAO is completed very rapidly after 156 min, as opposed to 208 min in the case of the conventional fully mixed AGS model. A similar observation is performed for the ammonium removal, which is also completed within 184 min by the Eawag AGS model, as opposed to 196 min predicted by fully mixed AGS model.

The differences observed in the nutrient profiles suggest modelling stratification over $H_{reactor}$ and plug-flow feeding influence in turn for the prediction of microbial activities (Figure 7). According to the fully mixed AGS model, the phosphate release rate (PRR) only starts to increase after 30 min of feeding once the nitrate uptake rate (NUR) is null, before reaching a constant rate of 10 mgP/L/h. In comparison, PRR starts directly with feeding and reaches a constant rate of 20 mgP/L/h when stratification/plug-flow feeding are predicted (Eawag AGS model). Similarly, the predicted phosphate uptake rate (PUR) is also strongly impacted by the model structure. A low PUR of max. 6 mgP/L/h is predicted by the fully mixed AGS model, as opposed to more than 30 mgP/L/h predicted by the Eawag AGS model. Comparable observations are possible regarding the ammonium and nitrate uptake rates (AUR and NUR, respectively), with slightly higher values predicted when stratification/plug-flow is considered in the model structure.

3.3.2. Implications for the microbial community composition

Modelling plug-flow feeding and sludge bed stratification has a major influence on the prediction of the sludge composition (flocs vs. granules) and on the microbial community composition (Figure 8). Fully mixed conditions favour the growth of OHO in the flocs and at the surface of granules (Figure 8). OHO growth predicted into the flocs and at the granules' surface by the fully mixed model exceed by 20 and 30% respectively predictions by the Eawag AGS model. The competitive OHO growth under fully mixed composition limits in turn the growth of PAO and GAO. The PAO + GAO concentrations are reduced by 13, 25, 25 and 11% in the layers #1–#4 of the granules of the fully mixed AGS model, as compared to the Eawag AGS model.

3.4. Functionalities of the Eawag AGS model (scenario #3)

3.4.1. Predicting system performances, e.g., effluent quality

An essential feature of the Eawag AGS model is its ability to predict the operation and performance of full-scale AGS plants. An example is given in Figure 9 for the dynamic of effluent solid concentration during bottom feeding with simultaneous

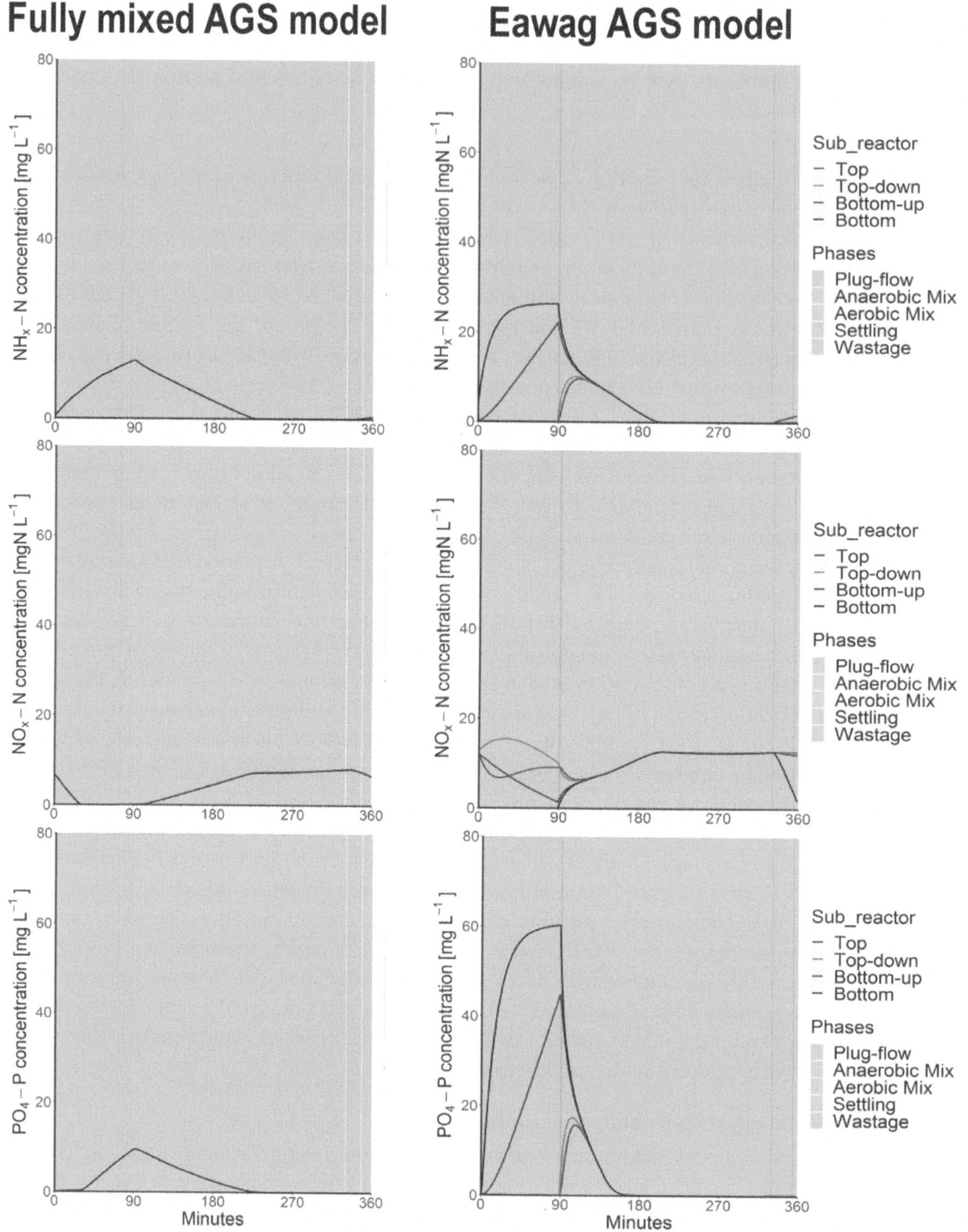

Figure 6 | Bulk concentration profiles of NH$_x$-N (top row), NO$_x$-N (intermediary row), and PO$_4$-P (bottom row) predicted by the fully mixed AGS model (left column) and Eawag AGS model (right column) over the different phases of the SBR cycle.

discharge of the treated wastewater at the top. Based on experimental data, the effluent solid concentration typically follows a decreasing trend during feeding, e.g., from 30 to less than 15 mgTSS/L after 40 min of feeding in the effluent. Such a decreasing trend is well predicted by our model with default settling parameters (no calibration). A gradual decrease of the effluent TSS concentration from around 20 mgTSS/L to less than 10 mgTSS/L at the end of the feeding is indeed predicted. Also, no leakage of ammonium from the influent into the effluent is predicted (perfect plug-flow, Figure 6). Ortho-phosphates are

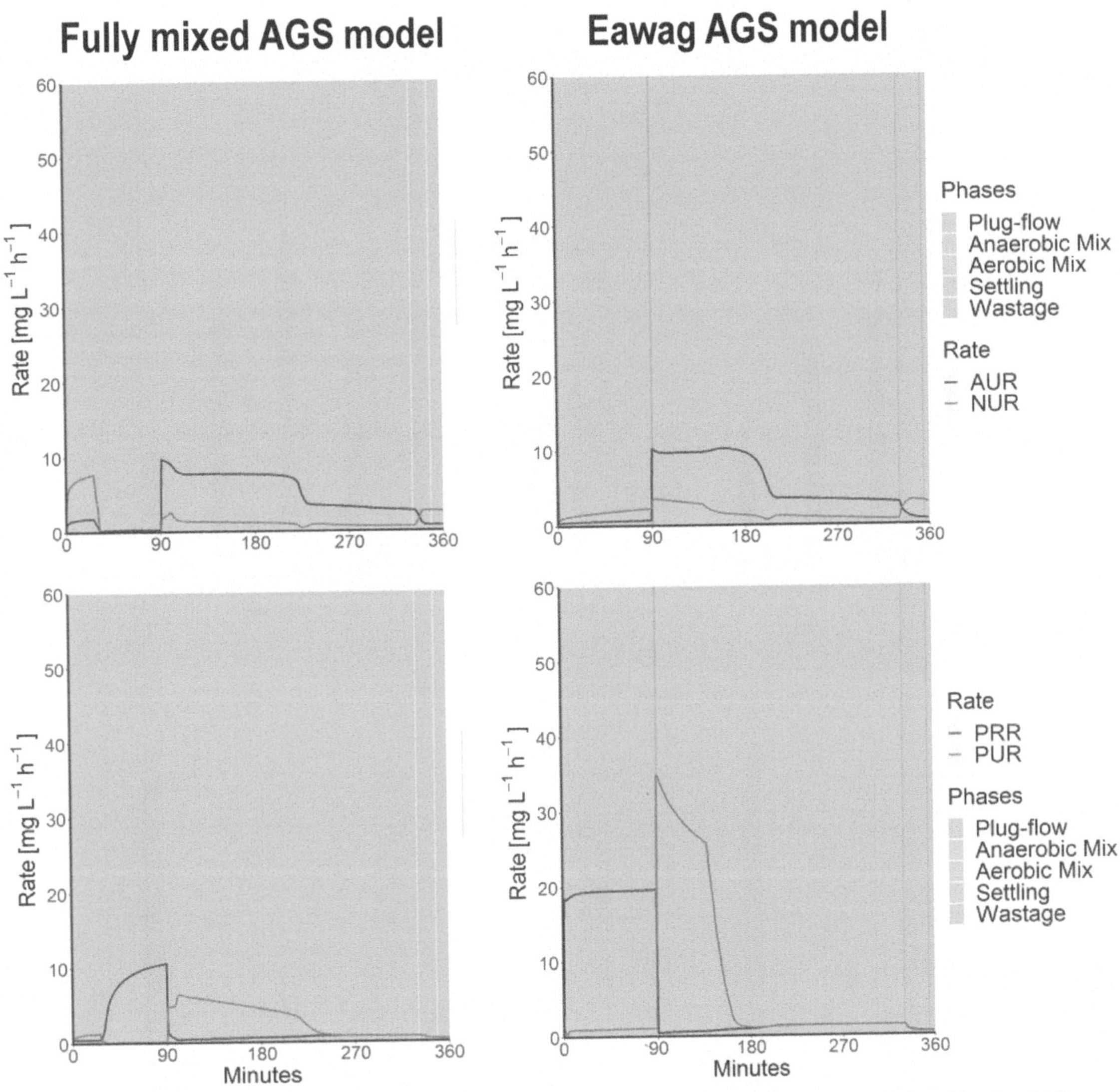

Figure 7 | Change in the nutrient conversion removal rates for the (a) conventional fully-mixed AGS model and for the (b) Eawag AGS model: ammonium uptake rate (AUR), NOx uptake rate (NUR), phosphate release rate (prr) and phosphate uptake rate (PUR).

released by both the granules in the bottom sub-reactor and the flocs in the bottom-up sub-reactor. The nitrate concentration follows a similar trend: with a slight decrease during the first half of the feeding followed by a slight increase and stabilisation during the second half.

3.4.2. Optimizing AGS-SBR operation using the Eawag AGS model

The Eawag AGS model also represents a tool to optimize the operation and performances of AGS systems, as for example the total nitrogen (TN) removal through smart aeration control (Figure 10). The Eawag AGS model allows predicting the penetration of oxygen over the depth of the granules (first row), the formation of the redox zone over the granule depth (middle row), and the resulting NOx concentration in the bulk (bottom row). At a dissolved oxygen set-point (DOSP) of 2 mgO$_2$/L in the bulk, granules are first partially penetrated by the oxygen, resulting in the formation of an aerobic zone at their surface while anoxic conditions prevail in their core (Figure 10, top left plot). Oxygen penetration gradually increased from 50 μm depth (90–140 min) to 100 μm depth (after 180 min). After 200 min, oxygen fully penetrates the granules, preventing the formation of anoxic conditions. Anoxic conditions thus form only for a short period in the core of the granule when a constant DOSP of 2 mgO$_2$/L is maintained (middle left plot), thus preventing denitrification and resulting in a rather large NOx accumulation in the bulk (bottom left plot).

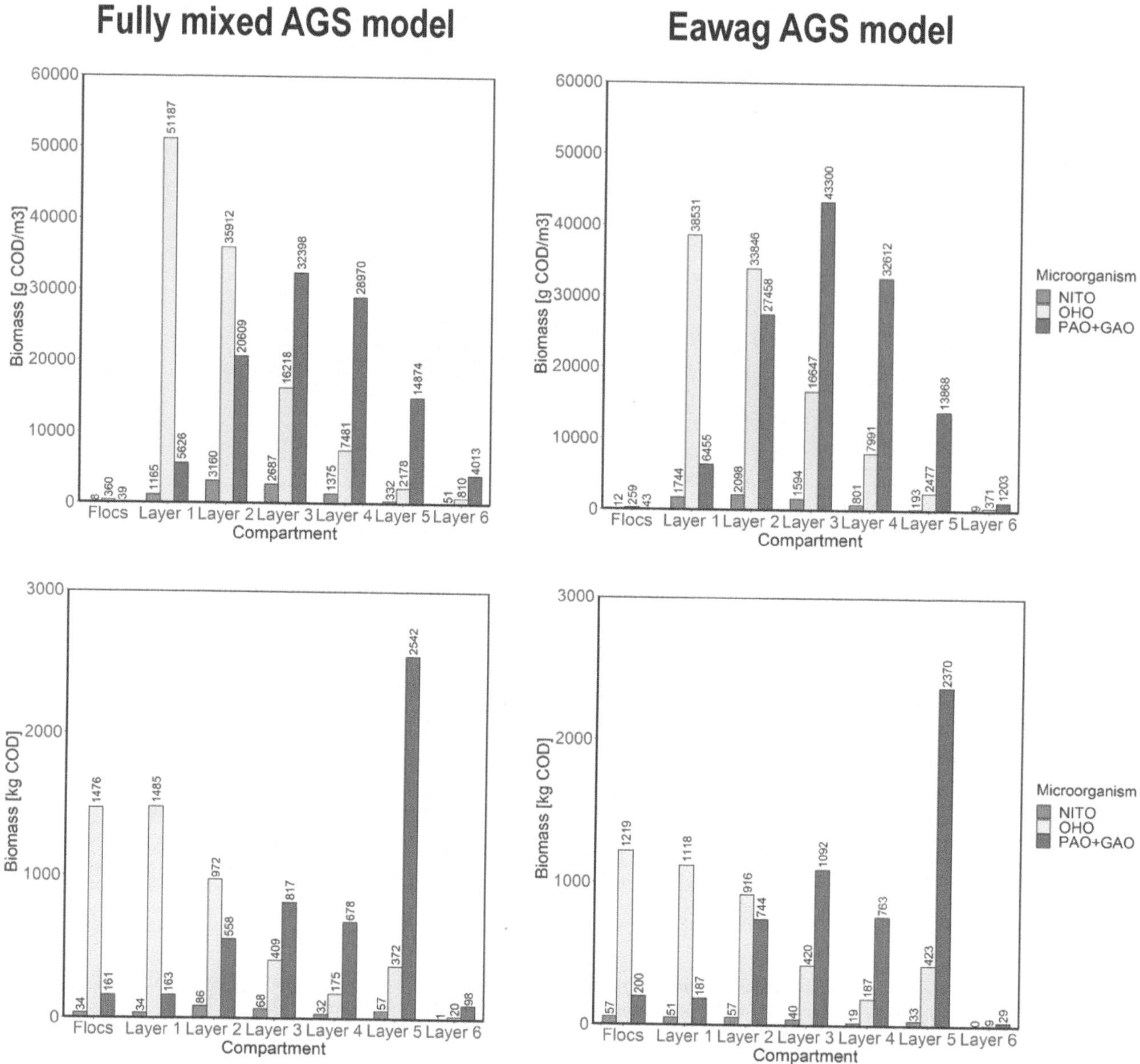

Figure 8 | Prediction of the microbial community composition within the flocs and granules, according to a conventional fully mixed AGS model (left column) vs. the Eawag AGS model (right column). Top row: biomass concentrations. Bottom row: biomass inventory.

The Eawag AGS model can be used by engineers to explore optimization strategies at a minimal cost/time, e.g., how to control aeration of their AGS-SBR to improve TN removal. In the example shown in Figure 10, a control of aeration at 2-DOSP (2 mgO$_2$/L for 30 min then 0.5 mgO$_2$/L) helped to better control the formation of the different redox zones and the utilisation of the organic substrate for denitrification (Figure 10, top right plot). As a consequence, and despite the complex composition of the influent (with a significant amount of non-diffusible electron donor), full nitrification-denitrification is predicted, resulting in a full TN removal (bottom right plot).

4. DISCUSSION

4.1. Gradients over H$_{reactor}$ matter: predict them!

Experimental evidence at laboratory and full-scale has long demonstrated that plug-flow feeding at the reactor bottom and selective removal of flocs is a key for the formation of granules. It is then intuitive to think that those processes should be

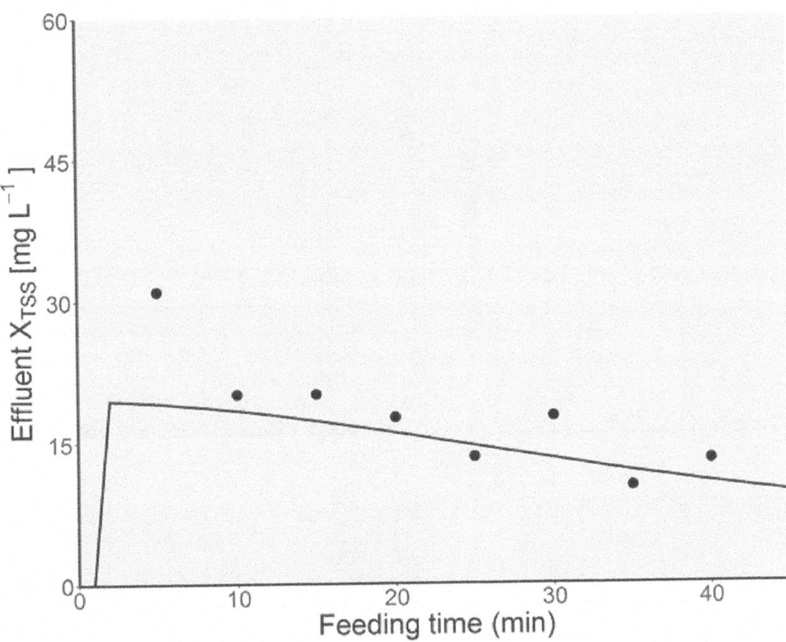

Figure 9 | Change in the TSS concentration in the effluent of AGS-SBR: experimental data from field test at Sarneraatal WWTP (black points) and predictions of the Eawag AGS model (black line).

incorporated into the structure of AGS models, while keeping the model structure 'as simple as possible, and as complex as needed' (Wanner *et al.* 2006). A first objective of our work was therefore to demonstrate that the integration of those processes into the structure of AGS models is *needed*. In other words, a key objective was to evaluate to what extent simulation results (biomass inventory, microbial activities and effluent quality) are affected by such additional model complexity. If model predictions are strongly impacted by plug-flow feeding/selective sludge removal and if those processes are essential features of the operation of full-scale AGS plant, it would then imply that those mechanisms must be considered in the structure of AGS models.

Our simulations confirm modelling gradients over $H_{reactor}$ does influence the prediction of microbial selection, and in turn the microbial activities and system performance (Figures 6–8). Modelling of the bed stratification and plug-flow feeding affects both the predictions of the biomass inventory and their spatial distribution over the granules radius. Fully mixed conditions during feeding favour accumulation of OHO, especially in the flocs and in the first layer of the granule (Figure 8). Ultimately, OHO and storing-organisms (PAO + GAO) were present in equal proportions within the granules (Figure 8). In contrast to this, modelling plug-flow feeding favoured accumulation of PAO + GAO, which in turn dominated the biomass inventory over OHO (Figure 8). Additionally, the concentrations of storing-organisms in the upper layers predicted by the Eawag AGS model exceeded the ones predicted by the fully-mixed model. The selection of slow growing microorganisms is a key for granulation and is enhanced during anaerobic feeding at the reactor bottom (de Kreuk & van Loosdrecht 2004; de Kreuk *et al.* 2005). The Eawag AGS model does not predict granule formation (constant total granule number/volume, one single size class) and is mostly dedicated to the modelling of plants with mature granules of steady diameter (Su *et al.* 2013). Yet, the competitive advantage predicted for PAO + GAO (inventory and spatial distribution) indirectly supports that granulation is favoured by a gradient over $H_{reactor}$, and that such processes must be predicted by AGS models.

As a result of the different spatial distribution and biomass inventories, microbial activities were also affected by the model structure, especially the biological phosphorus removal. The phosphate release rate predicted by the Eawag AGS model was twice as large as the one predicted by the fully mixed AGS model, while phosphate uptake rates were 5 times larger (Figure 7). As a consequence, ortho-phosphates were very quickly removed from the bulk during the aerated phase, as predicted by the Eawag AGS model. Our simulations were performed at a fixed SBR cycle duration (6 h total), and ultimately both models predicted full microbial conversion of substrates and nutrients. However, in practice, the length of the aerated phase is often controlled based on the ammonium bulk concentration. For example, the aerated phase is usually stopped once the

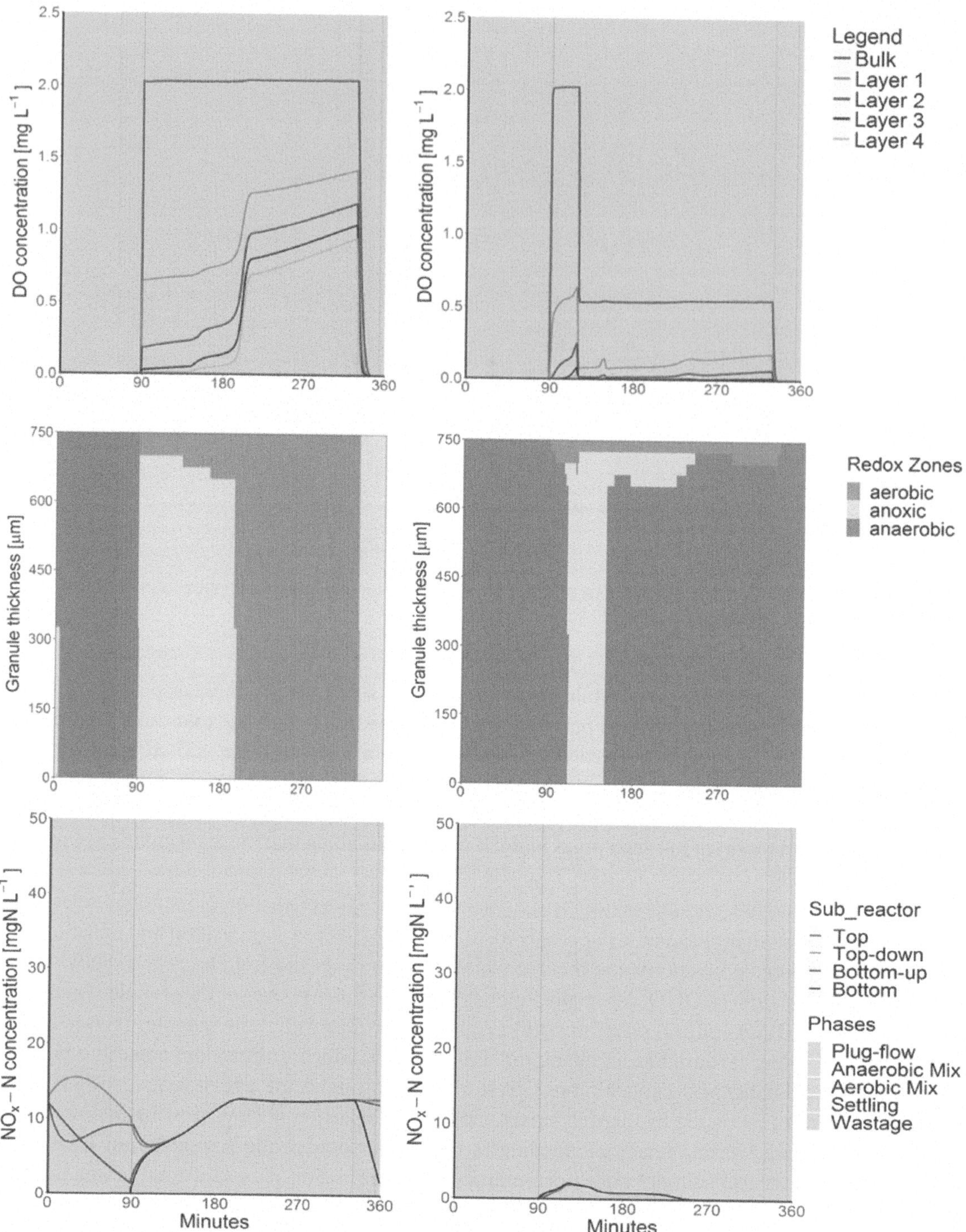

Figure 10 | optimization of the aeration control to improve total nitrogen removal (Scenario #3): aeration controlled at a constant dissolved oxygen set-point (DOSP) of 2 mg$_{O_2}$/L in the bulk (left-hand column) as opposed to 2-DOSP (first 2 mg$_{O_2}$/L and then 0.5 mg$_{O_2}$/L) (right-hand column row). Plots show the dissolved oxygen (DO) concentrations in the different layers (top row), the resulting redox conditions developing within the granules (middle row) and the NOx concentrations (bottom row). The distinction between aerobic, anoxic and anaerobic redox conditions is based on the half-saturation constants of growth on O$_2$ and NOx of OHO of the Sumo1 biokinetic model.

ammonium concentration measured in the bulk reaches a threshold value lower than 2 mgN/L. Modelling of gradient over H$_{reactor}$ slightly affected the AUR predicted by both models and full ammonium removal was completed at almost the same time by the two models (208 vs 184 min for the fully mixed and Eawag AGS model, respectively). But prediction of the AUR is important to identify when the aerated phase should ideally be stopped. If aerated phase had been stopped once an ammonium concentration below 2 mgN/L was reached (Swiss legal requirement), the legal requirements on phosphorus would not be met according to the predictions of the fully mixed AGS model (e.g., PO_4^{3-} concentration of 1.4 mgP/L vs 0.8 mgP/L as requirement for TP). This is explained by the much lower PUR predicted by the fully mixed AGS model, due to the competitive advantage of OHO over PAO that resulted from the fully mixed feeding conditions. In addition to being inappropriate to predict granules formation based on microbial selection of PAO + GAO, fully mixed AGS models are also inept to model system performances and to optimize SBR operation. Despite that, our model was not calibrated, comparing its predictions to the ones of the fully-mixed AGS model clearly demonstrate the importance of integrating the hydraulic selection into the structure of AGS models.

4.2. How to model gradients over $H_{reactor}$?

The Eawag AGS model combines a reactor model with a 1-D granule model (see section 4.3 for the rationale of such a choice). A key question was therefore what is the best approach to model gradients over $H_{reactor}$.

Dold *et al.* (2018) combined a 1-D biofilm model with a 1-D layered solids flux model. The 1-D layered solids flux model is used to predict the settling of mixed liquor solids (non-granules) over n-layers of equal depth during settling, while the granules settle immediately at the reactor bottom at the beginning of the feeding (Dold *et al.* 2018). Conceptually, the overall reactor model used by Dold *et al.* (2018) thus consists of two child-units for a total of $n + 1$ compartments: (1) a bottom child-unit containing the granules only and (2) a top sub-unit containing the supernatant with non-granule solids, divided in n-layers. An advantage of this approach is in offering a fine resolution for predicting gradients of growth conditions over the sludge bed, assuming a large n number is selected by the user.

In our model, gradients over $H_{reactor}$ are modelled using a series of four CSTRs combined with a correction code for plug-flow condition and using distinct settling models for flocs/granules. Our model thus resembles the model developed by Dold *et al.* (2018) with $n + 1 = 4$ compartments. The number of compartments in our model implementation is fixed, and equals the number of CSTRs connected in series. Are four compartments sufficient to capture the behaviour of AGS systems, or do we need a finer resolution on the gradient above the granules bed? In our model, the depth of each compartment is 'set' by the user based on full-scale data to reproduce a representative sludge bed stratification, characterized by granules at the bottom, flocs on the top of the granules, and supernatant above (divided into two layers). Our model proved to be effective to predict the sludge bed stratification and in turn the effluent solids (both concentration and dynamic over feeding time) (Figure 9). It is then not evident that a finer resolution above the settled granule bed, as in the Dold *et al.* model, is required (as long as effluent quality and plug-flow conditions are well predicted).

Also, what counts when modelling full-scale AGS systems is what happens at the reactor bottom during settling/feeding, i.e., in the granule bed. The model developed by Dold *et al.* (2018) assumes that granules immediately form a settled bed under non-mixed conditions, i.e., that their settling at the reactor bottom is 'forced', thus consistently resulting in the formation of an ideal bed stratification. This is a major difference to our modelling approach. Our model is in fact predicting the settling of granules and flocs and thus the resulting sludge bed stratification. The predicted stratification of the sludge bed thus results from a balance between the settling properties of flocs/granules and the operating conditions (settling time, feeding flow). These processes can result in the formation of an ideal (only granules at the bottom receive substrate) or not so ideal sludge bed (granules are pushed in the second CSTR where they compete with flocs for substrate). One advantage of our model is that it allows the user to test different operating conditions (e.g., varying feeding velocities) to explore their effect on the sludge bed stratification and resulting microbial competition between storing- and ordinary heterotrophic biomass.

One may also argue that hydrodynamic conditions are not 'perfect' in a real AGS-SBR due to some upward/downward mixing during the anaerobic feeding. One may question to what extent representing 'perfect' plug-flow conditions is therefore necessary. But predicting a 'perfect' plug-flow feeding actually helps predicting a full-uptake of diffusible organic substrates by storing-microorganisms (Figure 8), as observed on full-scale AGS systems (Pronk *et al.* 2015a, 2015b). Such conditions are key for achieving a successful granulation. In this sense, it is therefore acceptable to predict perfect plug-flow conditions during the anaerobic feeding like with the Eawag AGS model.

4.3. To model or not to model concentrations gradients over $Z_{granules}$?

A core component of the Eawag AGS model is its 1-D biofilm model. A fair question is *why* a 1-D model, and *why not* a 0-D model for example? Diffusion limits transport of solutes inside granules and results in the formation of concentration gradients over its radius (de Kreuk *et al.* 2010). Gradients of growth conditions govern in turn the spatial distribution of the microbial populations and ultimately the system performances (de Kreuk *et al.* 2007). Operation of full-scale AGS systems in sequencing batch mode also implies solute concentrations in the bulk (and thus diffusion) vary significantly over the cycle length (Layer *et al.* 2020b). The longer the anoxic zone exists within the granules, the longer denitrification takes place and the higher the N-removal efficiency (de Kreuk *et al.* 2007; Layer *et al.* 2020b). While diffusion is key to the performances of biofilm systems and can be easily described mathematically, some AGS models do not incorporate this phenomenon directly. Instead, the limitation of microbial conversion rates by diffusion is lumped into the apparent half-saturation constant values (Lübken *et al.* 2005; Baeten *et al.* 2019). But as a first step, identification of apparent half-saturation constants actually requires knowing the spatial distribution of the microbial populations, and therefore using 1-D biofilm models (Baeten *et al.* 2019). Also, apparent half-saturation constants are sensitive to changes in the operating conditions (e.g., DOSP) as well as growth conditions (influent composition, temperature) (Baeten *et al.* 2019), while such long-term variations are typical of full-scale WWTP operation. Therefore, the use of 1-D biofilm models that predict directly diffusion and spatial re-arrangement of the microbial populations appears more appropriate. 1-D biofilm models were developed in the 1980s (Wanner & Gujer 1985, 1986) and are commonly used today in engineering practice for the modelling of biofilm reactors (Boltz *et al.* 2010). Guidelines on how to calibrate and apply a biofilm model are now available (Rittmann *et al.* 2018) to help users gaining accurate and meaningful results. The use of a 1-D biofilm model for the Eawag AGS model was therefore justified, with regards to the goals of the model.

5. CONCLUSIONS

- A stratified AGS model was developed to predict the performances of aerobic granular sludge systems. This new AGS reactor model includes key features of full-scale AGS systems: (1) simultaneous fill-draw mode operation, (2) selective sludge removal, (3) coexistence of flocs and granules, (4) distinct settling models for flocs and granules. It allows predicting concentration gradients over $H_{reactor}$ and $Z_{granules}$ during the different phases of the SBR operation, selective removal of slow-settling biomass, dynamic of effluent quality, etc.
- While most existing AGS models disregard gradients over $H_{reactor}$, we demonstrate predicting those gradients has a non-negligible effect on the predictions of AGS models. Concentration gradients over $H_{reactor}$ during settling/plug-flow feeding impact the predictions of microbial selection, resulting microbial activities and ultimately effluent quality.
- Concentration gradients over $H_{reactor}$ can be accurately predicted with a series of four CSTR combined with plug-flow correction code. The height/volume of each CSTR can be adjusted based on knowledge from full-scale AGS plants, to reproduce a representative sludge bed stratification during non-mixed conditions.
- The Eawag AGS model is a valuable tool for both science and engineering practice. The Eawag AGS model can be used by scientists to better understand fundamental mechanisms or identify possibly important research gaps. The Eawag AGS model can also be used by engineers for the planning, design, optimisation, and evaluation of existing or future AGS-based plants.

ACKNOWLEDGEMENTS

The authors would like to express their strong gratitude to the following persons who supported this study: (1) Andrea Giesen and Sjoerd Kerstens from RoyalHaskoningDHV for providing access to AGS samples and for their expert advice, (2) Andreas Proesl (Wabag Water Technology Ltd) for providing access to Sarneraatal WWTP and for his expert advices, (3) Stefan Kälin (Sarneraatal WWTP) for providing access to the plant and support during field testing.

DATA AVAILABILITY STATEMENT

All relevant data are available from an online repository or repositories: https://opendata.eawag.ch/dataset/eawag-ags-model-package.

CONFLICT OF INTEREST

The authors declare there is no conflict.

REFERENCES

Baeten, J. E., van Loosdrecht, M. C. M. & Volcke, E. I. P. 2019 Modelling aerobic granular sludge reactors through apparent half-saturation coefficients. *Water Research* **148**, 556–556.

Beun, J. J., Heijnen, J. J. & van Loosdrecht, M. C. M. 2001 N-removal in a granular sludge sequencing batch airlift reactor. *Biotechnology and Bioengineering* **75** (1), 82–92.

Boltz, J. P., Morgenroth, E. & Sen, D. 2010 Mathematical modelling of biofilms and biofilm reactors for engineering design. *Water Science and Technology* **62** (8), 1821–1836.

Boltz, J. P., Johnson, B. R., Takacs, I., Daigger, G. T., Morgenroth, E., Brockmann, D., Kovacs, R., Calhoun, J. M., Choubert, J. M. & Derlon, N. 2017 Biofilm carrier migration model describes reactor performance. *Water Science and Technology* **75** (12), 2818–2828.

Campo, R., Sguanci, S., Caffaz, S., Mazzoli, L., Ramazzotti, M., Lubello, C. & Lotti, T. 2020 Efficient carbon, nitrogen and phosphorus removal from low C/N real domestic wastewater with aerobic granular sludge. *Bioresource Technology* **305**, 122961.

de Kreuk, M. K. & van Loosdrecht, M. C. M. 2004 Selection of slow growing organisms as a means for improving aerobic granular sludge stability. *Water Science and Technology* **49** (11–12), 9–17.

de Kreuk, M., Heijnen, J. J. & van Loosdrecht, M. C. M. 2005 Simultaneous COD, nitrogen, and phosphate removal by aerobic granular sludge. *Biotechnology and Bioengineering* **90** (6), 761–769.

de Kreuk, M. K., Picioreanu, C., Hosseini, M., Xavier, J. B. & van Loosdrecht, M. C. M. 2007 Kinetic model of a granular sludge SBR: influences on nutrient removal. *Biotechnology and Bioengineering* **97** (4), 801–815.

de Kreuk, M. K., Kishida, N., Tsuneda, S. & van Loosdrecht, M. C. M. 2010 Behavior of polymeric substrates in an aerobic granular sludge system. *Water Research* **44** (20), 5929–5938.

Derlon, N., Wagner, J., da Costa, R. H. R. & Morgenroth, E. 2016 Formation of aerobic granules for the treatment of real and low-strength municipal wastewater using a sequencing batch reactor operated at constant volume. *Water Research* **105**, 341–350.

Dold, P., Bill, A., Burger, G., Fairlamb, M., Conidi, D., Bye, C. & Du, W. 2018 *Modeling Full-Scale Granular Sludge Sequencing Tank Performance*. WEFTEC, New Orleans, LA, USA.

Kagawa, Y., Tahata, J., Kishida, N., Matsumoto, S., Picioreanu, C., van Loosdrecht, M. C. M. & Tsuneda, S. 2015 Modeling the nutrient removal process in aerobic granular sludge system by coupling the reactor- and granule-scale models. *Biotechnology and Bioengineering* **112** (1), 53–64.

Layer, M., Adler, A., Reynaert, E., Hernandez, A., Pagni, M., Morgenroth, E., Holliger, C. & Derlon, N. 2019 Organic substrate diffusibility governs microbial community composition, nutrient removal performance and kinetics of granulation of aerobic granular sludge. *Water Research X* **4**, 100033.

Layer, M., Bock, K., Ranzinger, F., Horn, H., Morgenroth, E. & Derlon, N. 2020a Particulate substrate retention in plug-flow and fully-mixed conditions during operation of aerobic granular sludge systems. *Water Research X* **9**, 100075.

Layer, M., Garcia Villodres, M., Hernandez, A., Reynaert, E., Morgenroth, E. & Derlon, N. 2020b Limited simulatenous nitrification-denitrification (SND) in aerobic granular sludge systems treating municipal wastewater: mechanisms and practical implications. *Water Research X* **7**, 100048.

Layer, M., Brison, A., Villodres, M. G., Stähle, M., Házi, F., Takács, I., Morgenroth, E. & Derlon, N. 2022 Microbial conversion pathways of particulate organic substrate conversion in aerobic granular sludge systems: limited anaerobic conversion and the essential role of flocs. *Environmental Science: Water Research & Technology* **8** (6), 1236–1251.

Lübken, M., Schwarzenbeck, N., Wichern, M., Wilderer, P., 2005 Modelling nutrient removal of an aerobic granular sludge lab-scale SBR using ASM3. In: *Aerobic Granular Sludge. Edition: Water and Environmental Management Series* (Bathe, S., de Kreuk, M. K., McSwain, B. S. & Schwarzenbeck, N., eds). IWA Publishing, London, UK.

MWH 2012 *Water Treatment: Principles and Design*. John Wiley & Sons, Hoboken, NJ, USA.

Ni, B. H. 2013 *Formation, Characterization and Mathematical Modeling of the Aerobic Granular Sludge. PhD Thesis*.

Ni, B. J. & Yu, H. Q. 2010 Mathematical modeling of aerobic granular sludge: a review. *Biotechnology Advances* **28** (6), 895–909.

Pronk, M., Abbas, B., Al-zuhairy, S. H. K., Kraan, R., Kleerebezem, R. & van Loosdrecht, M. C. M. 2015a Effect and behaviour of different substrates in relation to the formation of aerobic granular sludge. *Applied Microbiology and Biotechnology* **99** (12), 5257–5268.

Pronk, M., de Kreuk, M. K., de Bruin, B., Kamminga, P., Kleerebezem, R. & van Loosdrecht, M. C. M. 2015b Full scale performance of the aerobic granular sludge process for sewage treatment. *Water Research* **84**, 207–217.

Rittmann, B. E., Boltz, J. P., Brockmann, D., Daigger, G. T., Morgenroth, E., Sorensen, K. H., Takacs, I., van Loosdrecht, M. & Vanrolleghem, P. A. 2018 A framework for good biofilm reactor modeling practice (GBRMP). *Water Science and Technology* **77** (5), 1149–1164.

Shinya, M., Mayu, K., Goro, S., Akihiko, T., Yoshiteru, A., Satoshi, T., Cristian, P. & M, V. L. M. C. 2010 Microbial community structure in autotrophic nitrifying granules characterized by experimental and simulation analyses. *Environmental Microbiology* **12** (1), 192–206.

Su, K. Z. & Yu, H. Q. 2006 A generalized model for aerobic granule-based sequencing batch reactor. 1. Model development. *Environmental Science & Technology* **40** (15), 4703–4708.

Su, K. Z., Ni, B. J. & Yu, H. Q. 2013 Modeling and optimization of granulation process of activated sludge in sequencing batch reactors. *Biotechnology and Bioengineering* **110** (5), 1312–1322.

Takács, I., Patry, G. G. & Nolasco, D. 1991 A dynamic model of the clarification-thickening process. *Water Research* **25** (10), 1263–1271.

van Dijk, E. J. H., Pronk, M. & van Loosdrecht, M. C. M. 2018 Controlling effluent suspended solids in the aerobic granular sludge process. *Water Research* **147**, 50–59.

van Dijk, E. J. H., Pronk, M. & van Loosdrecht, M. C. M. 2020 A settling model for full-scale aerobic granular sludge. *Water Research* **186**, 116135.

Varga, E., Hauduc, H., Barnard, J., Dunlap, P., Jimenez, J., Menniti, A., Schauer, P., Vazquez, C. M. L., Gu, A. Z., Sperandio, M. & Takacs, I. 2018 Recent advances in bio-P modelling – a new approach verified by full-scale observations. *Water Science and Technology* **78** (10), 2119–2130.

Wanner, O. & Gujer, W. 1985 Competition in biofilms. *Water Science and Technology* **17** (2–3), 27–44.

Wanner, O. & Gujer, W. 1986 A multispecies biofilm model. *Biotechnology and Bioengineering* **28** (3), 314–328.

Wanner, O., Eberl, H., Morgenroth, E., Noguera, D., Picioreanu, C., Rittmann, B. & van Loosdrecht, M. C. M. 2006 Mathematical modelling of biofilms. IWA Scientific and Technical Report No.18, IWA Publishing, London, UK.

Weissbrodt, D. G., Holliger, C. & Morgenroth, E. 2017 Modeling hydraulic transport and anaerobic uptake by PAOs and GAOs during wastewater feeding in EBPR granular sludge reactors. *Biotechnology and Bioengineering* **114** (8), 1688–1702.

Xavier, J. B., de Kreuk, M. K., Picioreanu, C. & van Loosdrecht, M. C. M. 2007 Multi-scale individual-based model of microbial and bioconversion dynamics in aerobic granular sludge. *Environmental Science & Technology* **41** (18), 6410–6417.

First received 30 January 2022; accepted in revised form 13 July 2022. Available online 21 July 2022

doi: 10.2166/wst.2022.131

Mainstream short-cut N removal modelling: current status and perspectives

Gamze Kirim [a,b,*], Kester McCullough [c,d], Thiago Bressani-Ribeiro [e], Carlos Domingo-Félez [f], Haoran Duan [g], Ahmed Al-Omari[h], Haydee De Clippeleir [i], Jose Jimenez [h], Stephanie Klaus [d], Mojolaoluwa Ladipo-Obasa [i,j], Mohamad-Javad Mehrani [k,n], Pusker Regmi[h], Elena Torfs [l,m], Eveline I. P. Volcke [e,l] and Peter A. Vanrolleghem [a,b]

a modelEAU, Université Laval, 1065 avenue de la Médecine, Québec, QC G1 V 0A6, Canada
b CentrEau, Quebec Water Research Centre, 1065 avenue de la Médecine, Québec, QC G1 V 0A6, Canada
c School of Civil and Environmental Engineering, Cornell University, Ithaca, NY 14853, USA
d Hampton Roads Sanitation District, 1434 Air Rail Ave., Virginia Beach, VA 23455, USA
e BioCo Research Group, Department of Green Chemistry and Technology, Ghent University, Coupure Links 653, Gent 9000, Belgium
f Department of Environmental Engineering, Technical University of Denmark, Kongens Lyngby 2800, Denmark
g Australian Centre for Water and Environmental Biotechnology, The University of Queensland, Brisbane, QLD 4072, Australia
h Brown and Caldwell, 1725 Duke St. Suite 250, Alexandria, VA 22314, USA
i DC Water and Sewer Authority, 5000 Overlook Ave., SW., Washington, DC 20032, USA
j Department of Civil & Environmental Engineering, The George Washington University, 800 22nd Street NW, Washington, DC 20037, USA
k Faculty of Civil and Environmental Engineering, Gdansk University of Technology, Ul. Narutowicza 11/12, Gdansk 80-233, Poland
l Centre for Advanced Process Technology for Urban Resource recovery (CAPTURE), Frieda Saeysstraat 1, Gent 9000, Belgium
m BIOMATH, Faculty of Bioscience Engineering, Ghent University, Coupure Links 653, Gent 9000, Belgium
n Department of Urban Water and Waste Management, University of Duisburg-Essen, Universitätsstraße 15, 45141, Essen, Germany
*Corresponding author. E-mail: gamze.kirim.1@ulaval.ca

GK, 0000-0001-6964-4371; KM, 0000-0003-2637-8047; TB-R, 0000-0002-7363-7497; CD-F, 0000-0003-3677-8597; HD, 0000-0001-6679-3240; HD-C, 0000-0003-0541-8935; JJ, 0000-0003-3926-8779; SK, 0000-0003-4058-8104; ML, 0000-0001-6335-6897; M-JM, 0000-0003-2462-585X; ET, 0000-0002-5629-6950; EIPV, 0000-0002-7664-7033; PAV, 0000-0003-1695-1313

ABSTRACT

This work gives an overview of the state-of-the-art in modelling of short-cut processes for nitrogen removal in mainstream wastewater treatment and presents future perspectives for directing research efforts in line with the needs of practice. The modelling status for deammonification (i.e., anammox-based) and nitrite-shunt processes is presented with its challenges and limitations. The importance of mathematical models for considering N_2O emissions in the design and operation of short-cut nitrogen removal processes is considered as well. Modelling goals and potential benefits are presented and the needs for new and more advanced approaches are identified. Overall, this contribution presents how existing and future mathematical models can accelerate successful full-scale mainstream short-cut nitrogen removal applications.

Key words: anammox, deammonification, energy optimization, mathematical modelling, partial denitrification, partial nitration, resource optimization

HIGHLIGHTS

- Models for mainstream short-cut N removal processes are reviewed by considering their current practice and limitations.
- Modelling goals and potential benefits are presented from a modeller perspective to facilitate successful applications.
- More advanced modelling approaches are presented to overcome the addressed challenges and limitations.

GRAPHICAL ABSTRACT

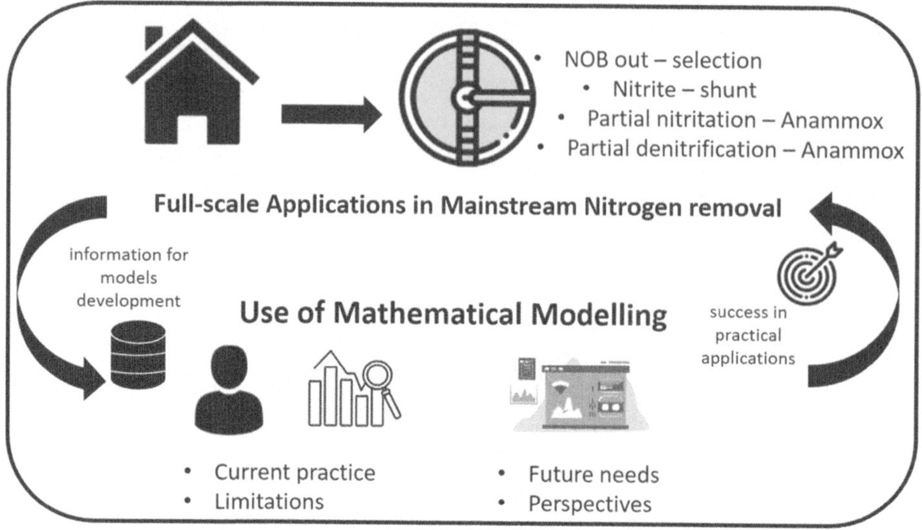

1. INTRODUCTION

Conventional nitrogen (N) removal by nitrification/denitrification is an energy and resource-intensive process: nitrification requires oxygen and alkalinity, and denitrification requires carbon as either influent carbon or supplemental carbon. Compared to nitrification-denitrification, the application of short-cut N removal processes in mainstream wastewater treatment has significant potential to save energy (oxygen demand), resources (carbon demand), and to pursue energy independence for water resource recovery facilities (WRRF). For that reason, short-cut N removal (deammonification and nitrite shunt) has received considerable attention over the last decade from both academia and practice.

Deammonification short-cuts the conventional N removal pathway by directly converting ammonium (NH_4^+-N) to nitrogen gas (N_2) via nitrite (NO_2^--N). The process relies on preventing the oxidation of nitrite to nitrate (NO_3^--N) and making nitrite available for anammox (Zhang *et al.* 2019). The availability of nitrite can be achieved through two pathways: partial nitritation-anammox (PNA) or partial denitrification-anammox (PdNA) (Figure 1). In practice, PNA and PdNA can be combined in full-scale applications for desired N removal performance.

The efficiency of deammonification has been proven for ammonium-rich wastewater such as treatment of side-streams resulting from dewatering of digested sludge, leachate, or industrial wastewaters (van Dongen *et al.* 2001; Wyffels *et al.* 2004; Volcke *et al.* 2005; Wett 2007; Ganigué *et al.* 2009; Lackner *et al.* 2014). Short-cut N removal can also be combined with a pretreatment process for carbon diversion in mainstream applications such as high-rate activated sludge (HRAS), chemically enhanced primary treatment and energy recovery in the side-stream, or even direct anaerobic sewage treatment (Kartal *et al.* 2010; Leal *et al.* 2016). This provides WRRFs with an excellent opportunity to move to energy-neutral or energy-positive operations (Jetten *et al.* 1997; Siegrist *et al.* 2008). However, full-scale applications are currently limited to side-stream treatment and only a few successful mainstream applications are reported so far (O'Shaughnessy 2016; Cao *et al.* 2017; Klaus *et al.* 2020).

PNA is a fully autotrophic process that consists of partial oxidation of NH_4^+-N to NO_2^--N (nitritation) and the anammox process in which NH_4^+-N is oxidized using NO_2^--N as an electron acceptor under anaerobic conditions without the need for carbon (Kartal *et al.* 2010) (Figure 1). Thus, the process requires the cooperation of ammonia-oxidizing bacteria (AOB) and anammox bacteria (AnAOB), and out-selection of nitrite-oxidizing bacteria (NOB). The successful application can reduce the required oxygen input by 60%, eliminate the carbon source demand and reduce the sludge production by 90% in comparison to conventional N-removal (Morales *et al.* 2015; Miao *et al.* 2016).

Partial nitritation-denitritation (nitrite-shunt) relies on partial nitritation of NH_4^+-N into NO_2^--N as the first step, then denitritation of NO_2^--N into N_2 as the second step by heterotrophic bacteria (HB). Thus, it consumes 25% less oxygen than complete nitrification and reduces the organic carbon demand by 40% compared to the full denitrification (Daigger 2014).

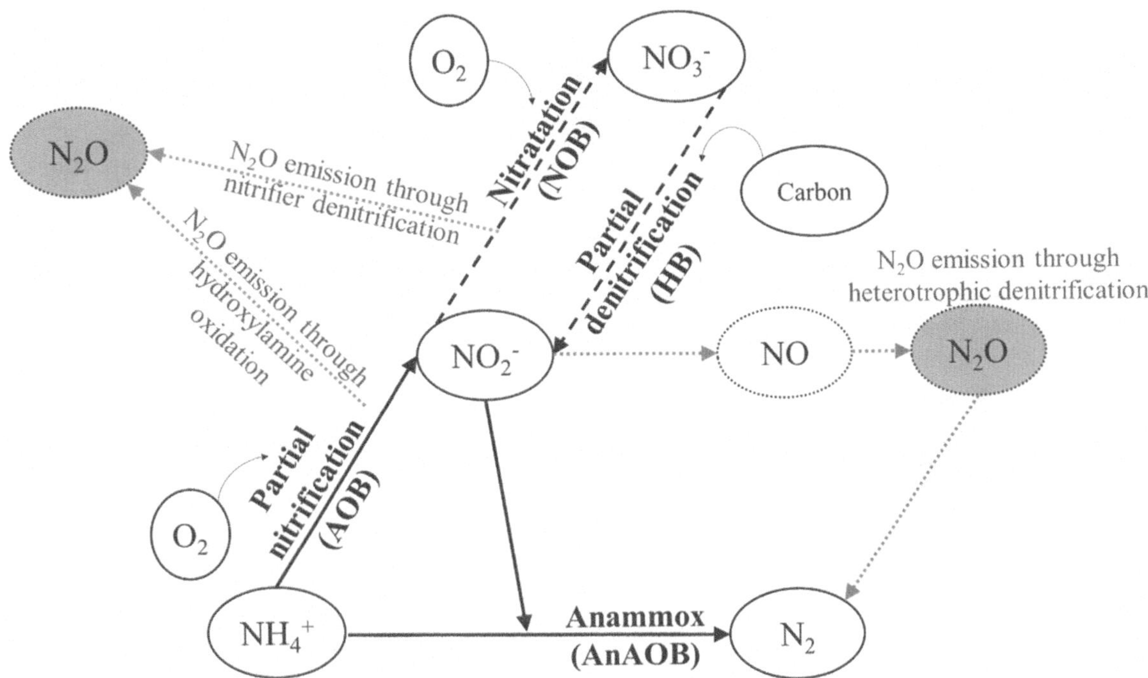

Figure 1 | Deammonification through partial nitrification-anammox and partial denitrification-anammox pathways including potential nitrous oxide emission pathways.

The process requires both out-selection of NOB and also carbon availability for denitritation. The nitrite-shunt process has been successfully implemented in side-stream treatment (e.g. SHARON process (Mulder *et al.* 2001)) and there is an interest to implement it in mainstream treatment in a single reactor (Jimenez *et al.* 2020). However, it is not commonly applied due to the lack of complete understanding of the underlying mechanisms for NOB out-selection and carbon availability for denitritation under these conditions.

During PdNA a portion of the influent NH_4^+-N is aerobically oxidized to NO_3^--N. The NO_3^--N is subsequently reduced to NO_2^--N via heterotrophic denitratation and the resulting mix of residual NH_4^+-N and NO_2^--N serves as substrate for the anammox process (Le *et al.* 2019a; Lu *et al.* 2021b) (Figure 1). This process does not require NOB out-selection and needs a carbon donor to achieve partial denitrification (Zhang *et al.* 2019). The PdNA process consumes slightly more resources (energy for aeration and carbon) than the PNA route; however, the nitrite generating pathway is understood compared to the NOB out-selection pathway (B. Ma *et al.* 2017; Lu *et al.* 2021a). Theoretically, 50% of aeration needs and 80% of carbon demand can be saved and sludge production can be reduced by 60% compared to conventional N removal (Z. Zhang *et al.* 2020). To overcome the external carbon need, recent research efforts aim to take advantage of the slowly biodegradable organics in wastewater or to produce soluble microbial products through fermentation (Ji *et al.* 2020; Liu *et al.* 2022). Also, simultaneous nitrogen and phosphorus removal with lower carbon and oxygen demand can be accomplished by combining endogenous partial denitrification with denitrifying phosphorus removal (X. Wang *et al.* 2019).

Current mainstream deammonification implementations include a variety of processes in laboratory and pilot scales including suspended, attached growth, or hybrid systems in single-stage or 2-stage reactors (e.g. Hoekstra *et al.* 2018; Klaus 2019; Le *et al.* 2019a; Huang *et al.* 2020). Due to the slow growth rate of anammox bacteria (Valverde Pérez *et al.* 2016; Lotti *et al.* 2015), an anammox retention mechanism is required to allow for adequate solids retention time (SRT). In a single-stage process, all biokinetic reactions occur in one basin which decreases both the investment and the operational costs (Pérez *et al.* 2014). Biofilm processes (Lotti *et al.* 2015; Gustavsson *et al.* 2020) or hybrid systems that combine suspended sludge with the biofilm systems such as integrated fixed-film activated sludge system (IFAS) (Cao *et al.* 2017) are used in these single-stage systems (W.-J. Ma *et al.* 2020) to retain AnAOB. In 2-stage systems, partial nitritation or full-nitrification (aerobic environment) and the deammonification processes (anoxic environment) occur in separate basins. Suspended or biofilm processes can be used in the aerated basin and biofilm-based processes may be used in the anammox basin (Regmi *et al.* 2014; Pérez *et al.* 2015).

Despite all the efforts, the mainstream application is still facing challenges due to NOB out-selection, wastewater characteristics, temperature and meeting the strict effluent criteria under dynamic loads (Cao *et al.* 2017). The competition for substrates and growth space between the different functional species is another major application challenge. High influent C/N ratios in the raw wastewater promote the growth of HB in the system, thus hampering deammonification under limiting oxygen concentrations (Gao & Xiang 2021). Also, lower ammonium concentrations and temperature variations make the stable out-selection of NOB very difficult in the PNA systems and lead to competition over oxygen by AOB, NOB and HB (M. Zhang *et al.* 2020).

Strategies are being developed to accelerate successful deammonification process implementation. Microorganisms involved in the short-cut N-removal processes are sensitive to operational and environmental conditions such as pH, dissolved oxygen (DO) level, temperature, SRT and the presence of inhibitors. Several control strategies have been adopted such as low or high DO operation, aerobic SRT, real-time aeration or oxidation-reduction potential control to take advantage of the growth characteristics and the kinetics difference between the microorganisms (Liu *et al.* 2020; Gao & Xiang 2021). However, the shift and adaptation of microbial communities' growth characteristics to mainstream conditions remain a challenge (Agrawal *et al.* 2018; Gao & Xiang 2021). In addition, due to high nitrite accumulation and ammonia conversion rates, the short-cut processes inevitably generate nitrous oxide (N_2O) as a by-product which is one of the most significant greenhouse gases (Castro-Barros *et al.* 2016; Li *et al.* 2020) (Figure 1). Further research is needed under mainstream conditions to achieve long-term process stability (e.g. varying loads or temperatures) and to understand the role of influent characteristics, varying substrates, intrinsic kinetics of the microorganisms involved and competition between them.

Mathematical models and model-based control strategies are under development to overcome implementation challenges and to deal with the complexity of mainstream deammonification (Agrawal *et al.* 2018). Through modelling, it is possible to identify the proper conditions for microbial competition under different operational and environmental conditions and to optimize the processes and implement deammonification successfully (Pérez *et al.* 2014; Liu *et al.* 2017; Shourjeh *et al.* 2021). However, the mechanistic models that are currently being used for the modelling are not sufficiently accurate to model short-cut N removal processes and require specific attention. For example, while the models include the key microbial groups, they do not consider the individual species which are crucial to reflect the competition among them and predict a community shift (e.g. NOB community shift (Liu & Wang 2013)). Also, different process configurations such as biofilm systems require specific sub-models such as the mass transport between the bulk liquid and the microorganisms inside the biofilm (Arnaldos *et al.* 2015; Baeten *et al.* 2019). Thus, the pilot and full-scale applications reported provide invaluable information for models development and to overcome bottlenecks while modelling efforts accelerate the success of practical applications.

The main focus of this paper is pointing out future needs and perspectives to facilitate the technology transfer between the model applications in research studies and accelerate the successful application of mainstream deammonification. The objectives of the presented review article are: (i) to illustrate the current practice in modelling of short-cut N removal processes, (ii) to reveal the challenges in modelling applications with examples, and (iii) to determine the immediate needs and the potential of more advanced modelling approaches and perspectives in the near future.

2. CURRENTLY APPLIED MODELLING APPROACHES AND LIMITATIONS

Modelling plays an essential role in improving process understanding and determining the optimal operating conditions and control strategies for applying short-cut nitrogen removal processes. There has been a concerted effort by academics and practitioners to model mainstream short-cut nitrogen removal recently, but these efforts have not yet resulted in consensus on process models and model parameter values that are transferable to practice. The activated sludge models (ASMs) developed by the IWA task group (Henze *et al.* 2006) are generally being used as default mathematical models for carbon and nutrient removal in wastewater treatment and are available in commercial software packages. For modelling short-cut nitrogen removal processes, ASMs have been used as well with extensions and modifications such as 2-step nitrification, 4-step denitrification, N_2O emission pathways and inhibition mechanisms. Despite its limitations, modelling can serve as a useful tool to improve process knowledge, screen technologies, and develop preliminary designs in a short-cut process. In this section, the required model selection mechanisms and current modelling perspectives for mainstream short-cut N removal processes are given.

2.1. NOB out-selection

Out-selection of NOB is crucial, especially for PNA or nitrite-shunt systems, and is widely recognized as one of the major challenges to mainstream application. The out-selection of NOB has been proven to be quite effective in warm nitrogen-rich wastewater streams (Lackner *et al.* 2014); due to the effect of elevated temperatures on growth rates for AOB and NOB (Hellinga *et al.* 1999), and also the high free ammonia (FA) and free nitrous acid (FNA) concentrations in side-stream liquors which inhibits the growth of NOB (Lackner & Agrawal 2015). However, FA inhibition is not possible in mainstream treatment due to the lower influent ammonium concentration (Cao *et al.* 2017). There are also reports of NOB out-selection achieved through side-stream generated FNA exposure (D. Wang *et al.* 2016) and alternating the sludge treatment strategy between FA and FNA can result in a stable nitrite-shunt with nitrite accumulation above 95% in the mainstream (Duan *et al.* 2019a).

The operating conditions to favour AOB and wash out NOB are thoroughly investigated in literature based on DO, pH, temperature and inhibitors. The intrinsic kinetics of these two groups of microorganisms including maximum growth rate and substrate half-saturations are crucial (Liu *et al.* 2020). DO affects the diversity and kinetics significantly, thus DO control to manipulate the competition for oxygen between AOB and NOB is one of the main strategies in mainstream conditions (Pérez *et al.* 2014; Jimenez *et al.* 2020). The oxygen half-saturation constant for AOB is generally accepted to be lower than for NOB which creates a disadvantage for NOB to compete for oxygen at low concentrations (Sin *et al.* 2008; Cao *et al.* 2017). On the other hand, the predominance of *Nitrobacter* or *Nitrospira*, which are the two main genera of NOB, affect the performance of NOB out-selection through DO control. The systems enriched with *Nitrospira* rather than *Nitrobacter* have a higher oxygen affinity, thus have lower oxygen half-saturation than AOBs and can be well adapted to low DO conditions (Regmi *et al.* 2014; Al-Omari *et al.* 2015). The use of transient anoxia is another approach by providing a lag-time for NOB to transition from anoxic to aerobic condition or nitrite limitation (Zekker *et al.* 2012; Gilbert *et al.* 2014). By consuming nitrite in anoxic conditions, heterotrophs restrict substrate availability for NOB in the aerobic phase (Regmi *et al.* 2014). Moreover, in mainstream treatment under limited DO, the AOB growth rate is higher than the NOB's at high temperatures (above 20°C) (Regmi *et al.* 2014; Yang *et al.* 2016). This allows operating the system at the SRT that is suitable for the growth of AOB and wash out the NOB (Blackburne *et al.* 2008). In addition, there are lab-scale works that support that NOB out-selection can be achieved at lower temperatures depending on the dominant NOB species in the system and the reactor configuration (De Clippeleir *et al.* 2013; Gilbert *et al.* 2015; Cao *et al.* 2017).

For modelling the short-cut processes, nitrite should be considered as an intermediary step in nitrification and denitrification. Modelling the two-step nitrification process is well established where NOB out-selection can be modelled through distinctly defined growth kinetics, substrate affinities, temperature and pH effect on AOB and NOB (Sin *et al.* 2008; Shourjeh *et al.* 2021). However, most simulation studies so far deal with side-stream conditions associated with high-strength nitrogenous wastewater where NOB out-selection can be achieved much more easily with direct pH and temperature effect on the NOB (Volcke *et al.* 2006, 2012; Van Hulle *et al.* 2007; Wett *et al.* 2010; Hubaux *et al.* 2015). For mainstream processes, community shifts and the changes in biokinetics become important which are not implemented in the ASM-based models yet. Favourable conditions to support the existence of AOB and facilitate NOB out-selection through different control systems could be demonstrated in the limited number of existing modelling works for mainstream treatment (Table 1). NOB out-selection is achievable in the models but there are still limitations to these models such as the calibrated half-saturation constants that can be a function of the environmental conditions, process configuration and operating conditions; thus, posing an issue of not being transferable to other systems.

2.2. Partial nitritation – denitritation

Partial nitritation-denitritation is an efficient biological pathway for N removal, but it has been challenged by the aforementioned difficulties (Section 2.1). The process can be applied as the first step of a 2-stage PNA process where nitrite-shunt is facilitated with controlled aerobic SRT. To improve effluent quality, the anammox process can be applied as a polishing step (Regmi *et al.* 2015a, 2015b; W. Zhang *et al.* 2020). Application of nitrite-shunt in mainstream treatment is desired through simultaneous nitrification denitrification because of the opportunity to enhance the utilization of the organic matter in the influent for denitrification. Note that the goal here is to maximize the use of the influent carbon through denitrification and not oxidation; hence, improving N removal while reducing energy consumption. However, it is not easy because the denitritation relies on utilizing the influent COD solely and thus the efficiency of the carbon pretreatment

Table 1 | Examples of modelling works for mainstream short-cut N removal processes

Modelled process	Reactor configuration	Modelling goal	Key findings	Limitations	Reference
Partial nitritation – denitritation	Lab-scale sequencing batch reactor (SBR)	Investigated the effect of aerobic duration on nitritation and NOB out-selection	AOB obtains more growth opportunity than the NOB which can occur only if the AOB reaction rate is higher than the NOB by considering the substrate concentrations	Simplified model excluding COD limitation, ammonification, assimilation of N	Blackburne *et al.* (2008)
	Lab-scale sequencing bench reactor (SBR)	Created an optimization framework and determined the optimal intermittent aeration profile to minimize energy consumption.	Rapid detection of the optimal aeration policies allowing an appropriate and prompt reaction to changes in the operation conditions in SBR processes	Comparing the results against previous publications due to the different conditions of the problem statement in each study	Bournazou *et al.* (2013)
	Pilot-scale activated sludge systems	Evaluated the performance of different process control strategies to achieve nitrite-shunt	The AvN strategy could improve the total nitrogen removal, sustain the NOB out-selection over the ABAC strategy and significant carbon savings could be achieved in comparison to conventional N-removal	Calibrating the model using dynamic input for pilot system due to the lack of a proper AVN controller in the model	Al-Omari *et al.* (2015)
	Biofilm system (pure modelling study)	Determine the influence of biofilm properties (e.g. water-biofilm interface thickness, substrate diffusivities) on NOB suppression	Increased biofilm thickness poses more resistance to diffusive transport of DO, thus limiting the NOB growth	Pure modelling study based on assumed influent characterization	Liu *et al.* (2020)
Partial nitritation – anammox	Granular sludge reactor (pure modelling study)	Investigated microbial community interactions at low temperatures and sensitive parameters leading to NOB repression.	The nitrite half-saturation coefficient of NOB and anammox bacteria proved non-influential on the model output. The maximum specific growth rate of anammox bacteria proved a sensitive process parameter.	Granule size distribution was not considered. Model excluded heterotrophic growth.	Pérez *et al.* (2014)
	Lab-scale SBR	Described the microbial interaction among ammonia-oxidizing archaea (AOA), AOB and anammox bacteria	AOA outcompete AOB under low ammonium concentration and low dissolved oxygen conditions, indicating a better partnership with anammox bacteria.	Oxygen inhibition coefficient for anammox derived from a marine species ($1\ \mathrm{g\ DO\ m^{-3}}$).	Pan *et al.* (2016)
	Granular sludge reactor (pure modelling study)	Evaluated control strategies to minimize the impact of influent disturbances, using dynamic model simulations.	Fixed or adaptive ammonium set point control strategy with DO limit enabled PNA.	Model assumed a homogeneous granule size.	Wu (2017)
	Lab-scale granular sludge reactor	Investigated the impacts of C/N ratio, DO concentration and granule size distribution on the process performance.	The granule size distribution should be incorporated in the model to accurately describe the granular anammox system.	External mass transfer boundary layer was not taken into account.	Liu *et al.* (2017)
	CSTR (PN)+ granular sludge reactor (anammox) (pure modelling study)	Investigated the effect of operational conditions on final effluent N concentration.	TN discharge standard of $10\ \mathrm{gN\ m^{-3}}$ is only met for temperatures above 25°C.	Model assumed a homogeneous granule size. Pure modelling study based on assumed influent characterization.	Bozileva *et al.* (2017)

(Continued.)

Table 1 | Continued

Modelled process	Reactor configuration	Modelling goal	Key findings	Limitations	Reference
Partial nitritation – anammox	MBBR and IFAS (pure modelling study)	Explored operating conditions in IFAS and MBBR systems.	IFAS can achieve higher nitrogen removal at lower airflow rate than MBBR. PN occurs mainly in the biofilm in MBBR and it is restricted to suspended solids in IFAS.	Pure modelling study based on assumed influent characterization. Steady-state simulations.	Tao & Hamouda (2019)
	Lab-scale SBR	Investigated how the composition of the flocs and the NOB concentration respond to changes in DO, fraction of flocs removed per cycle, and maximum volumetric anammox activity	Selective NOB wash out by controlling the DO-setpoint and/or the flocs removal allowed anammox bacteria to act as 'NO2-sink' in the biofilm.	The oxygen inhibition of anammox bacteria was not explicitly modelled. The biofilm was modelled as zero-dimensional, and spatial gradients were neglected. Perfect biomass segregation between flocs and biofilm.	Laureni *et al.* (2019)
	HRAS-PNA (pure modelling study)	Assessed the feasibility and long-term stability of the granular sludge PNA reactor through dynamic modelling and simulation.	Anammox as a dominant process for N removal. The HRAS-PNA system was more sensitive to temperature compared to the conventional activated sludge system.	Model assumed a homogeneous granule size. Pure modelling study based on assumed influent characterization (BSM2).	Jia *et al.* (2020)
	MBBR and IFAS (pure modelling study)	Assessed the role of external boundary layer resistance with respect to bacterial competition and nitrogen removal capacity, focusing on low temperatures (10°C).	The external mass transfer resistance promoted the metabolic coupling between anammox and ammonia oxidizing bacteria. The effectiveness of the nitrite sink depended on the anammox bacteria sensitivity to oxygen.	Steady-state simulations. Pure modelling study based on assumed influent ammonium concentration (without COD).	Pérez *et al.* (2020)
	Lab-scale SBR	Used bifurcation analysis to assess the co-existence of AOB and NOB and the ideal scenario where NOB is completely removed from the reactor.	Good process performance shown even under sub-optimal conditions (i.e., NOB remain in the reactor).	Pure modelling study using a novel mathematical analysis, based on experimental results from Laureni *et al.* (2019).	Wade & Wolkowicz (2021)
	Granular sludge reactor (pure modelling study)	Evaluated the impact of feeding disturbances on the performance of a single-stage PNA granular reactor.	A cascade control strategy based on DO manipulation to derive the ammonium set-point value proved efficient under dynamic influent conditions.	Pure modelling study based on assumed influent characterization (BSM1).	M. Zhang *et al.* (2020)
	UCT-MBR system (pure modelling study)	Comparatively assessed the anammox process and conventional heterotrophic denitrification in an existing UCT-MBR system.	Anammox process weakens the system's resilience to influent fluctuations.	Pure modelling study neglecting diffusion limitations on MBR systems.	Shao *et al.* (2021)
Partial denitrification-anammox	Pilot-scale MBBR, pilot-scale IFAS, & full-scale biological sand filter	Modify SUMO2 (based on ASM) to describe, parameterize, and calibrate partial denitrification for VFA and methanol substrates with and without the presence of AnAOB.	Nitrite preference needed to be removed from denitrification rates. Nitrate residual (via electron flow regulation) needed to be added to model, with additional parameters. Denitrification rate differentials and competition over nitrite with anammox were handled well by model kinetics.	New parameters in the model required batch-test calibration. New rate equations may not fully capture electron competition.	Al-Omari *et al.* (2021)

TN, total nitrogen; MBBR, moving bed biofilm reactor; PN, partial nitritation; UCT-MBR, University of Cape Town membrane bioreactor system; VFA, volatile fatty acids process. In addition, the mechanisms for achieving controllable simultaneous nitrification denitrification are not well understood yet (Klaus 2019; Jimenez *et al.* 2020) and it is out of the scope of this paper.

Operational strategies for stable nitrite-shunt performance have not been demonstrated yet in large-scale systems (Xu *et al.* 2017; H. Wang *et al.* 2019). Modelling of nitrite-shunt requires a holistic approach by simultaneously monitoring the influent dynamics and using the data for controlling the operational conditions and considering their effect on competition for the different substrates. Recent research efforts mostly deal with the application of deammonification in full-scale mainstream as opposed to nitrite-shunt (Table 1).

2.3. Partial nitritation – anammox

For application of PNA processes for mainstream treatment, a further concern is the competition between HB and anammox bacteria for nitrite. Few modelling studies under mainstream conditions have dealt with the interaction between AnAOB, AOB, NOB and ordinary heterotrophs (e.g., Al-Omari *et al.* 2015). As for side-stream applications, simulations showed that the availability of some influent COD can lower the effluent nitrate concentration (produced by anammox) by heterotrophic denitrification and thus increase the total nitrogen removal efficiency of anammox reactors (Hao & van Loosdrecht 2004; Mozumder *et al.* 2014). In many anammox studies, the presence of HB was ignored, and the COD in the reactor was neither measured nor considered in mass balances (Schielke-Jenni *et al.* 2015). Nevertheless, heterotrophic denitrifiers have been widely found in anammox reactors and can account for up to 23% of the biomass in biofilm reactors even without organic matter in the influent because HB could grow both on soluble microbial products and decay released substrate (Ni *et al.* 2012).

When modelling the anammox stoichiometry, one should be careful because the experimentally determined yield for the overall metabolic reaction ($Y^{Met'}_{X/NH_4}$=0.172 g COD/g NH_4^+-N, Strous *et al.* (1998)) mistakenly has been used in many simulation studies (Hao *et al.* 2002; Volcke *et al.* 2010; Ni *et al.* 2012). Ammonium in the anammox process in ASMs is consumed in both catabolic and anabolic reactions, while the yield coefficient only accounts for the ammonium taken up in the catabolic path. Therefore, the experimentally determined yield cannot be directly implemented. To remedy this, Jia *et al.* (2018) proposed an alternative stoichiometric equation based on the biomass yield per amount of ammonium consumed in the overall metabolic reaction or recalculating the widely model yield (Y^{mod}) from the measured $Y^{Met'}_{X/NH_4}$ as follows;

$$Y^{mod} = 1/1/Y^{Met'}_{X/NH_4} - i_{NXB}$$

where i_{NXB} represents the nitrogen content of anammox bacteria (g N g COD^{-1}).

Modelling outputs of the mainstream PNA process also tend to be strongly affected by the oxygen inhibition coefficient of anammox bacteria, as demonstrated by Pérez *et al.* (2020) for MBBR and IFAS systems. The external mass transfer resistance plays a significant role in this case, as the biofilm can be exposed to lower DO concentrations under thicker boundary layers. Although low oxygen levels benefit anammox bacteria, AOB activity could be hampered too. In this case, ammonia-oxidizing archaea (AOA) could be better coupled to anammox bacteria, as the latter can thrive at lower DO levels compared to AOB (You *et al.* 2009). Nevertheless, few models consider the contribution of AOA to nitrogen conversions (Pan *et al.* 2016).

Steady-state simulations have been used to assess PNA systems (Bozileva *et al.* 2017), and only a few studies address dynamic simulations (Table 1). Under varying influent conditions, a higher sensitivity to temperature oscillation was found compared to steady-state simulations (Jia *et al.* 2020). Therefore, dynamic simulations are recommended to assess process feasibility and long-term stability.

2.4. Partial denitrification – anammox

An early indication of PdNA process in the literature was in a 2-stage pilot-scale PNA system (Regmi *et al.* 2015b). Eliminating the challenge of NOB out-selection has led to significant research interest in the PdNA process (Du *et al.* 2017; Cao *et al.* 2019; Du *et al.* 2019; X. Wang *et al.* 2019; B. Ma *et al.* 2020; You *et al.* 2020; Chen *et al.* 2021) and multiple successful pilot-scale and full-scale implementations (McCullough *et al.* 2021). As complete nitrification is already well established in process models, the primary challenge of modelling PdNA is understanding the reduction of nitrate to nitrite under various process configurations, substrate concentrations, redox conditions, and with/without the presence of anammox.

PdNA is typically implemented as a 2-stage system which can be an integrated process (Le *et al.* 2019a; Li *et al.* 2019) or a post-polishing (Campolong *et al.* 2018). Controlling and maximizing denitratation over denitritation is key to PdNA start-up and performance because full denitrification results in a loss of carbon efficiency and no ammonia removal. Multiple carbon sources have been demonstrated to be effective for partial denitrification such as acetate, glycerol, or methanol (Campolong *et al.* 2018; Le *et al.* 2019b; McCullough *et al.* 2021), ethanol (Du *et al.* 2017), fermentation products (Cao *et al.* 2013; Ali *et al.* 2020) and endogenously stored carbon (Ji *et al.* 2017; X. Wang *et al.* 2019). Beyond carbon source selection, numerous factors can affect the partial denitrification efficiency including influent C/N ratio, pH, sludge retention time and microbial population. Recent literature suggests that under mainstream conditions, the dominant factor impacting partial denitrification efficiency is in fact the nitrate concentration in the reactor (nitrate residual) (Le *et al.* 2019a). Nitrate residual has been demonstrated to be an effective method for controlling PdNA and partial denitrification efficiency during start-up (Schoepflin *et al.* Manuscript in preparation), pilot-scale (Le *et al.* 2019a), and full-scale (McCullough *et al.* 2021). Since it can be well correlated with partial denitrification efficiency and was used successfully for process control, the nitrate residual is found to be a key parameter in modelling PdNA as well (Al-Omari *et al.* 2021).

Modelling PdNA requires accurate description of partial denitrification in the model and the ability to address microbial competition over electron donors and acceptors. While denitrification is included in ASM 1, ASM 2-2d and ASM3 models as a single-step (Henze *et al.* 2006), at least two-step denitrification must be modelled for PdNA. Monod functions were used to model two-step denitrification by Hellinga *et al.* (1999), in ASM3 by Ni & Yu (2008) and for granular sludge by De Kreuk *et al.* (2007). Many multi-step denitrification models included terms to inhibit denitratation in favour of denitritation (referred to here as 'nitrite preference') based on experimental observations where nitrite accumulation was not observed or was impeded. Wett & Rauch (2003) proposed a model using the ASM structure in which the rate of denitritation is elevated and the rate of denitratation is hampered by an inhibition factor based on the ratio of nitrate to nitrite. Similar practices appeared to be common practice, as Hiatt & Grady (2008) state that denitrification models from Gujer and colleagues included a nitrite inhibition term for each step of nitrogen reduction, e.g. Wild *et al.* (1995). Their four-step activated sludge model for nitrogen (ASMN) includes a nitric oxide inhibition factor for denitrification as well as separate anoxic reduction factors for each denitrification rate. More advanced three-step or four-step denitrification models introduce additional complexity and parameters, but do not appear to be necessary to successfully model PdNA. These models are discussed in more detail in Section 4.4.

The nitrite preference applied to all electron donor sources used for denitrification hindered the ability to induce partial denitrification in existing models. Al-Omari *et al.* (2021) introduced a partial denitrification switch based on bench-scale and pilot-scale observations where partial denitrification was observed in batch tests and was controlled in continuously fed systems by maintaining nitrate residual. This was observed when using specific carbon sources such as acetate and glycerol but not methanol. The 'nitrite preference' was eliminated in this case for the nitrate reduction reactions using the volatile fatty acids (VFA) substrate state variable.

2.5. N_2O emissions in short-cut N removal processes

N_2O is produced mainly from three microbial pathways: the NH_2OH oxidation and nitrifier denitrification pathways carried out by AOB, and the heterotrophic denitrification pathway carried out by HB (Duan *et al.* 2017) (Figure 1). N_2O is a potent greenhouse gas with a large global warming potential of 265 times that of CO_2, as well as the single most significant ozone-depleting substance (Edenhofer *et al.* 2014). Accurate assessment, modelling, and mitigation of N_2O emissions is critical due to the large contribution of N_2O to the WRRF carbon footprint (Vasilaki *et al.* 2019). Peng *et al.* (2020) analysed the carbon footprint of the mainstream PNA process, including electricity consumption, N_2O emissions, and reduction in carbon emissions through energy recovery. Based on this study, the N_2O emission factor for mainstream PNA should not exceed 0.78% to achieve a carbon-neutral operation. However, the N_2O emissions from the PNA process could be staggeringly high, with up to 10% of the nitrogen removed emitted as N_2O (Domingo-Félez *et al.* 2014; Staunton & Aitken 2015).

The capability to simulate N_2O emissions by mathematical models allows the consideration of N_2O emissions in the design and operation of short-cut nitrogen removal processes. N_2O models have evolved from simple one-pathway models (from AOB or from 3rd or 4th steps of heterotrophic denitrification) (Hiatt & Grady 2008; Mampaey *et al.* 2013; Pan *et al.* 2013) to two-pathway AOB models (Ni *et al.* 2014; Pocquet *et al.* 2016) that involve both N_2O production pathways from AOB. Thereafter, integrated pathway models were developed that consider both AOB and heterotrophic N_2O generations (Guo & Vanrolleghem 2014; Domingo-Félez & Smets 2016; Spérandio *et al.* 2016), and indicating that N_2O can also be further removed but only by HB (Guo & Vanrolleghem 2014). These integrated pathway models have been applied in conventional systems as a powerful tool to evaluate N_2O mitigation strategies (Chen *et al.* 2019; Duan *et al.* 2020, 2021).

With essentially the same principles of N_2O generation, models developed in nitrification and denitrification systems can be directly applied to the short-cut nitrogen removal processes. For example, Wan *et al.* (2019) integrated the two pathway N_2O model by Pocquet *et al.* (2016), with the heterotrophic N_2O production model by Hiatt & Grady (2008) to describe N_2O emissions in a single-stage PNA reactor. The established N_2O model was used to evaluate the effects of operational conditions on N_2O emissions from the PNA system and identified the potential operational conditions to reduce emissions.

Challenges and uncertainties are present in the modelling of N_2O emissions in mainstream short-cut nitrogen removal systems. The largest uncertainty results from the lack of datasets. N_2O emission data from full-scale mainstream short-cut nitrogen removal systems are limited. On the other hand, the data obtained from lab-scale applications show high emission ratios ($5.2 \pm 4.5\%$ of total inorganic N removed) (Roots *et al.* 2020). In addition, many short-cut nitrogen removal evaluations lack N_2O data. Therefore, adequate nitrogen species profiling data becomes crucial to establish nitrogen mass balances. Another challenge is that N_2O process models tend to overparameterize the description of biological pathways and this impacts the N_2O model calibration (Domingo-Félez & Smets 2020).

3. HOW MODELLING CAN ACCELERATE THE FULL-SCALE MAINSTREAM APPLICATIONS

In the mainstream application of short-cut N removal processes, mathematical models are being mostly used to simulate the behaviour of biological processes under different operating scenarios and cost-effectiveness (Shourjeh *et al.* 2021). Although currently experimental work in laboratory and pilot-scale systems prevails over mechanistic model analysis for short-cut processes (M. Zhang *et al.* 2019; Z. Zhang *et al.* 2020), modelling efforts can help to accelerate mainstream full-scale applications. In this section, several modelling goals and potential benefits are presented from a modeller perspective to reveal how modelling could be useful to achieve successful applications (Figure 2).

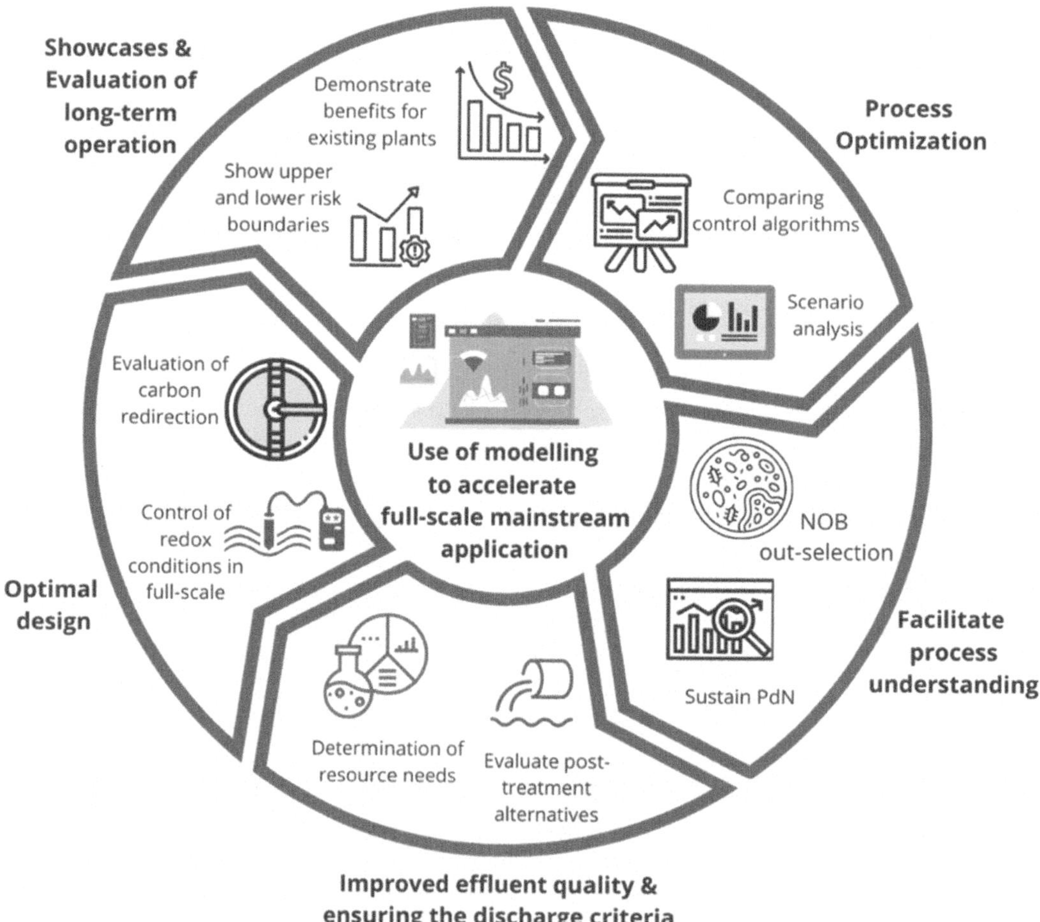

Figure 2 | How to benefit from modelling to accelerate full-scale mainstream application.

3.1. Facilitate process understanding

As presented in Section 2, models are mainly being used to facilitate the process understanding and improve process efficiency through improved operation and process control. Existing models can improve mainstream NOB out-selection applications for PNA processes since it allows assessment of the impacts of stable and dynamic influent conditions, biofilm characteristics and aeration control strategies (Al-Omari *et al.* 2015; Rosenthal *et al.* 2018). With the ability to simulate varying influent conditions such as low or high COD/N ratios, modellers can examine the hydrolysis of biodegradable substrates and impacts on processes (Tao & Hamouda 2019). In addition, models permit the exploration of how different biofilm systems (MBBR, granular sludge, IFAS and filters) would work for mainstream applications (Boltz *et al.* 2011; Rosenthal *et al.* 2018).

Similar to NOB out-selection, existing models can be used to improve the efficiency and nitrogen removal contribution via PdNA pathway by evaluating the impact of feedstock carbon characteristics, biofilm characteristics, process configuration, aeration controls and operational conditions.

3.2. Process optimization and control

Pilot and full-scale applications of different control algorithms are increasing rapidly to provide the operational and environmental conditions to sustain short-cut N removal in mainstream treatment. Modelling can be used to design control strategies to improve the operation of existing plants and serves as a time and cost-saving tool for the evaluation of different control algorithms (Salem *et al.* 2002; Gernaey *et al.* 2004). To fulfil the effluent quality standards, reduce carbon footprint and keep the operational costs at a low level, it is imperative to use control strategies to optimize resource consumption (Ostace *et al.* 2011; Revollar *et al.* 2020).

To achieve mainstream deammonification, several real-time aeration control strategies have been developed and applied such as continuously low DO operation, AvN (NH_4^+-N vs NO_x^--N) with intermittent aeration, ammonia-based aeration control (ABAC) (Regmi *et al.* 2014; Sadowski *et al.* 2015; Poot *et al.* 2016). Together with the real-time aeration control, the application of short aerobic SRT is a robust control strategy that relies on adjusting the wasting rate, aerobic volume and DO setpoints to suppress NOB in PNA systems (Regmi *et al.* 2014; Han *et al.* 2016). Nitrate-based COD dosing control is needed in PdNA systems to ensure nitrite availability for anammox (Le *et al.* 2019a). Finally, yet importantly, dynamic feedforward control of the AvN setpoint to respond to changes in the influent ammonia can be implemented to maximize the PdNA efficiency with savings in carbon dosage (Le *et al.* 2019c). By using the information and experience obtained from lab and pilot-scale studies, modelling can be used to compare features of different process control strategies, provide a proof of concept, and develop the control algorithms (e.g. feedforward control based on the influent dynamics or multi-objective control).

3.3. Evaluation of post-treatment requirement

Depending on the short-cut nitrogen removal process efficiency and the effluent requirements, the post-treatment might be needed for the removal of nitrate, nitrite or residual ammonia. Often in the short-cut N process, nitrate may still be present in the effluent, especially when NOB out-selection efficiency is compromised (Gustavsson *et al.* 2020; Muñoz 2020). This requires post-treatment such as heterotrophic denitrification with external carbon dosage. Effluent ammonia can be an issue at high loading rates and a reaeration zone might be needed. Models can be used to design and test the required post-treatment alternatives coupled with control strategies, determine the reactor volume needed, air consumption or the applicable type of chemical dosage. It can be helpful to ensure the effluent limits and analyse the overall resource consumption in long-term operation.

3.4. Demonstration of the potential benefits of technology add-ons for existing processes under real conditions

Modifying a conventional N removal process with short-cut N removal processes can reduce overall energy and resource consumption of a WRRF. To demonstrate and quantify the gains of such modifications, models can be used for the capacity analysis of a WRRF by identifying the SRT, volume necessity, chemical consumption and stoichiometric needs for the processes (Shao *et al.* 2021). For example, Al-Omari *et al.* (2015) used a calibrated model for a hypothetical scenario and demonstrated the potential of 60% external carbon savings by converting the conventional nitrification denitrification system to mainstream deammonification with the Blue Plains WRRF case study. Moreover, such evaluations provide an assessment of the potential performance to meet the total inorganic nitrogen effluent limits and associated risk in long-term operations. Demonstration of the benefits of technology for existing plants can be achieved when transferable validated process models are available, and the model parameter values are obtained.

Furthermore, from a plant-wide modelling perspective, mainstream short-cut N removal processes are interacting with other biochemical and chemical processes in the WRRF which would be important to capture both the dynamics of the plant and the potential environmental impacts (Arnell *et al.* 2017). For example, through plant-wide modelling, the influence of carbon redirection process on the downstream short-cut N removal process performance and the microbial species could be predicted; including the greenhouse gas emissions, carbon and energy footprint of the WRRF (Mannina *et al.* 2019). Simulating the whole plant would allow determining if the side-stream processes can be used to support the mainstream processes through seeding the bacteria (AOB or Anammox, e.g. Wett *et al.* 2015) or providing FNA or FN exposure (e.g. D. Wang *et al.* 2016; Duan *et al.* 2019a). In addition, combined biological P and short-cut N removal potential can be investigated. Thus, adopting the plant-wide modelling approach can be useful to demonstrate the gains in terms of process performance and stability in full-scale implementations.

4. THE NEED FOR NOVEL MODELLING APPROACHES

Despite its limitations, modelling is proven to be useful to help achieve full-scale mainstream N removal processes (Section 2&3). The applications at different scales provide invaluable information to support and improve the models while the modelling efforts can help to accelerate successful practical application and foresee challenges. Within this section, the needs and future perspectives on model applications are presented. Based on the discussion outputs of the *Workshop* of the *7th IWA/ WEF Water Resource Recovery Modelling Seminar Mainstream Short-cut Nitrogen Removal Modelling: From research to full-scale implementation, do we have what we need?*, a feasibility versus importance chart is given in Figure 3 for the future

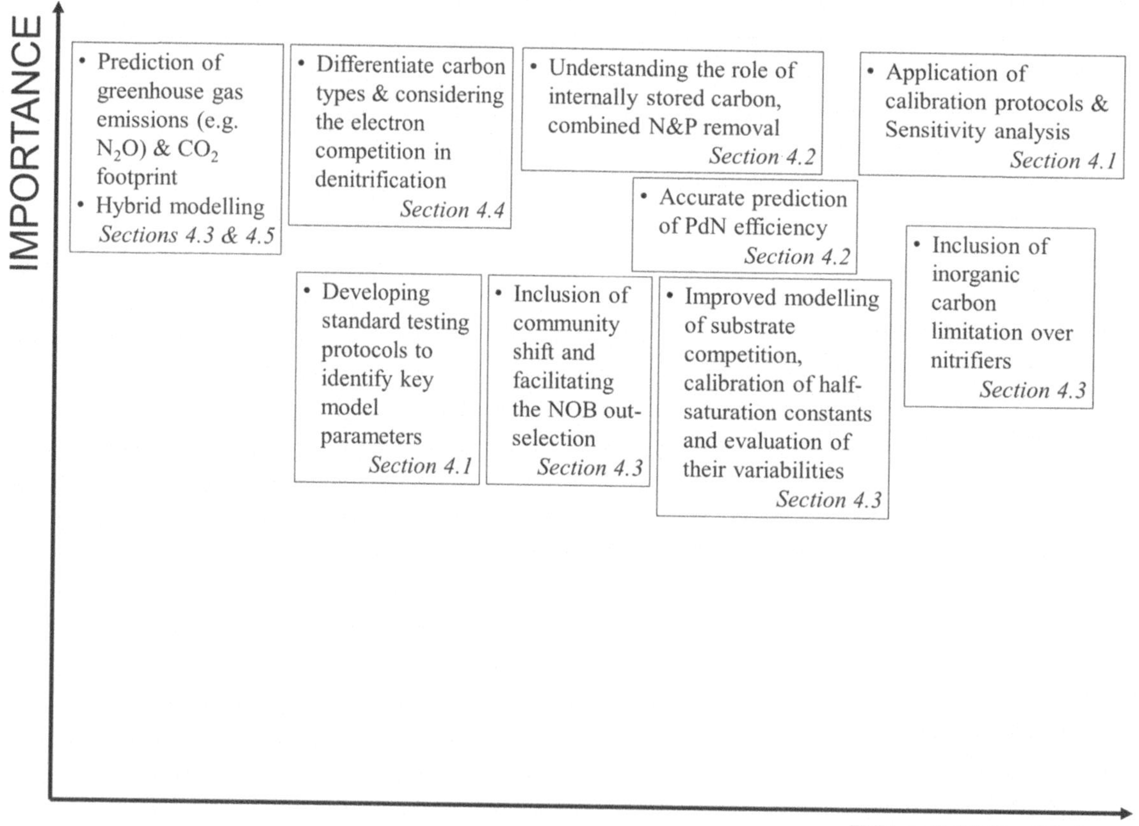

Figure 3 | Importance vs feasibility chart for the future works needed to improve short-cut N removal modelling.

works that are urgently needed (the discussion points with less importance are not included). The main subjects are discussed in detail below.

4.1. Parameter estimation methodology

For the modelling of short-cut processes, robust calibration methodologies and model developments are needed to find the optimal range of influential model parameters and improve process understanding and stability. Existing process models for short-cut nitrogen removal are commonly used for scenario analysis to evaluate process performance under varying operating conditions (e.g. Mozumder *et al.* 2014; Pérez *et al.* 2014). The models are mostly goal-oriented (i.e. applied for a specific purpose and case), but are seldom followed by experimental validation. For example, the degree of stratification of AOB and NOB in nitritating granules (Soler-Jofra *et al.* 2019), or the postulated pH-driven inhibition of NOB in a PNA system (Y. Ma *et al.* 2017) are later confirmed by pH microprofiles which is related to the NOB out-selection in biofilm systems (Ma *et al.* 2021).

Based on the analysis of the reported modelling studies in literature, it appears that the applied calibration methodologies are seldom following good modelling practice which makes the comparison between modelling studies difficult. This is probably due to the complexity and change of the microbial communities studied with lab-scale systems when they are applied to full-scale (Rieger *et al.* 2012). Selection of the parameters is typically based on expert knowledge following the parsimonious principle (i.e. keep the number of parameters to be estimated as small as possible), and the calibration methodology itself mostly is not well described (Al-Omari *et al.* 2015; Trojanowicz *et al.* 2019; Drewnowski *et al.* 2020; Roots *et al.* 2020). One should be aware that this type of model calibration may lead to unrealistic model outputs when the model is run without boundary conditions and is extrapolated for process optimization. Moreover, the model parameter values estimated in this way may not be transferable to other systems. Proper model validation can protect from such unrealistic extrapolation as it would indicate to what extent extrapolation is feasible.

The application of robust sensitivity analysis and the use of more rigorous calibration methodologies would make the comparison between model structures possible and the simulated model outputs would become reproducible and transferable. Importantly, it also enables the modeller to identify the main model factors to pay attention to in terms of model algorithms and the data (Mannina *et al.* 2010; Vangsgaard *et al.* 2013). This identification can reveal which data is more informative for model calibration, thus can guide the experimental efforts (e.g. the higher sensitivity of growth rates regarding performance within the biofilm vs bulk recommends *in situ* microprofiling data is more informative for calibration than the bulk measurements (Y. Ma *et al.* 2017)). This issue may be minimized if long-term consistent data sets can be obtained which is typically the weakest link in modelling of short-cut processes currently.

Sensitivity analysis is the basic method to identify the most appropriate parameter subsets to be calibrated. In short-cut process models, calibration coupled with the use of more advanced approaches has been presented but has remained limited (e.g. normalized sensitivity function (Baek & Kim 2013; Y. Ma *et al.* 2017)). Sensitivity analysis can be done in a simple way through local sensitivity analysis (LSA), for which the number of model simulations that need to be run is low, typically $2*N+1$ runs (N: number of parameters) (De Pauw & Vanrolleghem 2006a). However, the LSA approach does not allow identifying interacting factors, thus may miss some of the influential model parameters and limit the use of the calibrated model. On the other hand, global sensitivity analysis (GSA) provides information on how the model outputs are influenced by the variation on the model inputs and parameters over a wide range of possible parameter values (Cosenza *et al.* 2013). Thus, GSA provides an overall view of the sensitivity and determines the important model parameters much more reliably. In addition, GSA covers interacting parameters much better (Lackner & Smets 2012). Its computational demand is higher than the one of LSA, but it can be optimized by using an appropriate sampling method, e.g. Latin Hypercube Sampling (Vanrolleghem *et al.* 2015).

Sensitivity analysis will further guide the identification of the most influential model inputs and allow for the design of future monitoring campaigns for full-scale applications with the aid of model-based optimal experimental designs (OED) (Ledergerber *et al.* 2019). OED can help design the experimental studies that are highly informative and meaningful for calibration and validation of the models and thus will make models that reliably predict the model outputs (De Pauw & Vanrolleghem 2006b). Finally, standard testing methods/protocols for the measurement of key model parameters would ensure good initial estimates of their values can be assigned since such estimates allow reducing the calibration efforts and making different case studies comparable. Examples of such standard tests are activity tests for nitrification rates and measurements of half-saturation coefficients.

4.2. Role of complex carbon types and their effects on partial denitrification

While the PdNA model from Al-Omari *et al.* (2021) has been calibrated to a range of data from pilot-scale systems and started to be applied to full-scale processes, many questions remain that may require more advanced models (Section 2.4). The developed model can simulate partial denitrification only with methanol and VFA (glycerol or acetate) substrates but have not yet been tested for more complex carbon types. Other electron donors need to be better understood to model PdNA systems that rely on complex influent COD or fermentation products. Fermentation products have been shown to be useful for partial denitrification at pilot-scale (Ji & Chen 2010) and PdNA (Ali *et al.* 2020). Influent COD has also been used to drive partial denitrification, although this may require in-line fermentation (Shi *et al.* 2019; Ji *et al.* 2020; Liu *et al.* 2022) or pH elevation above 9.0 (Qian *et al.* 2019; Shi *et al.* 2019).

Additional work needs to be done to better understand intracellular carbon storage and its impact on PdNA. Simple carbon sources like glycerol and acetate can be stored by heterotrophs (Le *et al.* 2019b), and the resulting denitrification dynamics of storage versus immediate use is not well understood in these systems. Studies have shown that internally stored carbon from influent wastewater is a promising alternative for partial denitrification (Ji *et al.* 2017, 2018). Denitrifying glycogen-accumulating organisms present (dGAO) in enhanced biological phosphorus removal (EBPR) systems may also preferentially perform denitratation (Rubio-Rincón *et al.* 2017; H. Wang *et al.* 2019), leading to a potential synergy between next generation EBPR and PdNA systems.

4.3. Microbial competition modelling

Estimation of half-saturation constants: An efficient deammonification system must ensure both the activity of AOB and AnAOB while the growth of NOB is inhibited (especially in PNA systems). Also, the growth of HB should be controlled since it may shift the process from deammonification to conventional nitrification and denitrification (Cao *et al.* 2019; Gao & Xiang 2021). In low substrate availability conditions such as short-cut N removal processes, microbial competition for substrate becomes important for both application and modelling. The substrate competition might be between AOB and AnAOB for ammonium as electron donor; AOB, NOB and HB competition for oxygen as electron acceptor; and AOB and NOB competition for inorganic carbon (Shourjeh *et al.* 2021). Microorganism growth kinetics and substrate affinities provide a useful tool to understand and model the substrate competition and population shift. Expanding the previous models by accommodating for the competition between main species, we will have to revisit and further study the wide range of kinetic parameters for AOB, NOB, and AnAOB (Cao *et al.* 2017).

Impacts of influent load fluctuation, environmental and operational conditions on model accuracy becomes more important and perhaps more challenging for addressing competition more accurately in the model. On the other hand, strategies and operational conditions are being tested mainly in lab-scale and modelling efforts are often lacking consistent long-term experimental data (Table 1). For example, related to the successful NOB out-selection, the models are mostly applied for the data out of a short-term monitoring period and only reflects the operational conditions that favour the AOB or AnAOB activities while NOB is inhibited (Blackburne *et al.* 2008; Cui *et al.* 2017; Laureni *et al.* 2019; Al-Hazmi *et al.* 2022). Thus, not including the challenging operational conditions that promote NOB growth may lead to an inadequate calibration of NOB activity and their out-selection. Most of the models are calibrated to reflect the periods for stable operation; hence, the half-saturation values would not change drastically except the acclimation and adaptation conditions. Using long-term, where organisms are well adapted to the operational conditions, and consistent performance data for dynamic simulation with robust calibration methodologies could be useful to overcome the calibration challenges related to competing species in the short-cut process (e.g. Vangsgaard *et al.* 2013).

The growth kinetics of microorganisms is substrate-limited based on Monod's formulation used in the ASMs which states a fixed relation between growth rate and bulk substrate concentration (Henze *et al.* 2006). Thus, the substrate half-saturation constant (or affinity constant) is crucial. It represents the substrate concentration corresponding to a half-maximal rate of growth (Riet & Lans 2011). The lack of understanding and proper characterization of half-saturation constant variability between different systems leads to the need for frequent calibration of half-saturation constants. However, calibration may lump the effects of different phenomena in the system (such as mixing, advection and biofilm diffusion limitations) and that may lead to a variation on the apparent values of the half-saturation constants, thus affecting the prediction power of the calibrated model (Arnaldos *et al.* 2015). For example, the current practice for oxygen mass transfer modelling is to apply overly simplified models. These models require multiple assumptions that are not valid for most applications and are highly uncertain (e.g. α-factor). Thus, the calibration of DO half-saturation constants is affected especially in short-cut processes where small deviations in the simulated DO concentration have already a significant impact on the biological conversion processes (Amaral *et al.* 2019). In addition, transport of oxygen within the floc is driven by diffusion where temperature and DO gradient become important factors (Manser *et al.* 2005; Arnaldos *et al.* 2015). A high variation of such factors may require implementing the half-saturation constant as a model variable and calculating its value during the simulation (e.g. K_{NO3} dependence on biomass growth rates, thus the temperature by Shaw *et al.* 2013). Furthermore, the different species of the same genera of microorganisms may have different affinities to the same substrate. For example, *Nitrobacter* or *Nitrospira* are the two main genera of NOB and their predominance in the system is affected by nitrite concentration (Nogueira & Melo 2006). For these reasons, estimation of the half-saturation constants should be handled very carefully by considering the transport phenomena and the effect of biological limitations to model the substrate competition properly. Advanced modelling tools in combination with the biokinetic models such as computational fluid dynamics and population balance models could be useful in full-scale systems (Arnaldos *et al.* 2015). The inclusion of different species in the same genera with their substrate affinities might be considered especially to model NOB suppression (see below: Inclusion of community shift). Also, when the kinetics are under dual limitation, both substrates should be taken into account in the model (Al-Omari *et al.* 2015).

Alternatively, hybrid models are an interesting avenue to pursue in this domain. Hybrid models combine mechanistic models with data-driven techniques and as such leverage both process knowledge and the power of data analytics. As a result, hybrid models can have good extrapolation properties while being less laborious to develop than strictly mechanistic models. Hybrid models can be used in many different configurations and with many different combinations of data-driven tools. However, they have been described to be specifically powerful in situations where the overall process model structure is quite well defined but specific knowledge on variability of subprocesses is missing (von Stosch *et al.* 2014). In these situations, adding a data-driven component to the well-described mechanistic model structure can specifically learn the missing dynamics and compensate for the uncertainties in the mechanistic model which can be much more efficient than going through extensive laboratory experiments to develop in-depth understanding of all the factors contributing to the variability in some subprocesses. The modelling of kinetics in short-cut N removal processes falls exactly within this definition with a lot of process knowledge and corresponding process knowledge available but a lack of mechanistic description of all factors influencing the kinetics (e.g. mixing, biofilm diffusion limitations).

Inclusion of community shift (interspecies competition): It is well-known that at low oxygen concentrations, NOBs are outcompeted because of their relatively low affinity for oxygen. Experimental evidence is available suggesting different reactor operating conditions and control strategies favour the presence of different nitrifying species, which could be different ammonium oxidizers (Bougard *et al.* 2006) and/or different nitrite oxidizers (Dytczak *et al.* 2008). Considering the oxygen concentration as the key variable governing the population shift, Volcke *et al.* (2008) modelled the shift in AOB species in a biofilm reactor. But also NOB could adapt to control strategies by community shifts, resulting in failed NOB control and subsequently disrupting the PNA or nitrite-shunt process. With the low DO (between 0.16 and 0.37 mg O_2/L) NOB control strategy, NOB gradually developed competitive edges over AOB for oxygen uptake by shifting the NOB community to be dominated by *Nitrospira* (Liu & Wang 2013). Similarly, the NOB community could adapt to inactivation from free ammonia (FA) or free nitrous acid (FNA), by shifting to *Nitrobacter*, or *Nitrospira* respectively (Duan *et al.* 2019a, 2019b).

The adaptation of NOB by community shift poses a challenge to the modelling of mainstream PNA or nitrite-shunt processes. The current ASMs with 2-step nitrification could predict the out-selection of NOB under low DO conditions given the commonly higher oxygen affinity constant of NOB than that of AOB. However, as the NOB community adapted to the low DO condition by shifting the dominant NOB genus from *Nitrobacter* to *Nitrospira*, NOB exhibited an increasingly higher affinity to oxygen (lower half-saturation constant). Modelling such NOB activities in the mainstream PNA or nitrite pathway is not feasible with one set of parameters. The inclusion of detailed microbial diversity (interspecies competition) in models for process design and operation may be warranted in cases where it affects the macroscopic reactor performance (e.g. nitrite accumulation). Taking the low DO control scenario as an example, if the model included kinetics parameters/processes for both *Nitrospira* (lower oxygen half-saturation constant) and *Nitrobacter* (higher oxygen half-saturation constant), the model may be able to predict the community shift, and the adaptation of the NOB community to the low DO conditions under long-term operation. The inclusion of the community shift in the model may allow predicting a more reliable NOB control performance, and thus a more stable mainstream PNA or nitrite-shunt process. Note that if the model includes the kinetics parameters/processes for competing interspecies, a complete washout of one the species can be predicted at certain conditions and it would never reappear even if the conditions later shift towards more favourable conditions. However, in reality, this might happen seldom and the two species may always coexist, only switching their predominance depending on the conditions. The reason might be the seeding of NOB from the influent wastewater. It has been reported that different NOB species may be present in raw wastewater reaching the WRRFs (Jauffur *et al.* 2014; Yu *et al.* 2016). This could potentially facilitate the development of the NOB community, which may result in unstable NOB suppression in the mainstream (Duan *et al.* 2019b). To reflect this situation in the model, these interspecies must be included in the input of the model.

Inorganic carbon limitation of AOB: Inorganic carbon can be a crucial factor influencing conversions, as demonstrated by Bressani-Ribeiro *et al.* (2021) for the treatment of anaerobic effluents. Dynamics of AOB have been reported to change significantly under inorganic carbon limitation while the NOB activity remains stable (Guisasola *et al.* 2007; Ma *et al.* 2015; Zhang *et al.* 2016). Biesterfeld *et al.* (2003) showed that nitrification rates are affected by an inorganic carbon shortage (below 45 mg $CaCO_3$ L^{-1}) independently of pH. Instead of inorganic carbon, alkalinity (expressed as bicarbonate) is typically introduced in models to predict possible pH changes and close charge balances (Rieger *et al.* 2012). The ASMs follow that approach, in which alkalinity limitation on (single-step) nitrification is described with Monod kinetics (Henze *et al.* 2006).

The alkalinity limitation considered in these models focuses on unfavourable pH conditions rather than inorganic carbon limitation. Nevertheless, modelling approaches considering inorganic carbon limitation effects on autotrophs have been proposed, described with Monod or sigmoidal kinetics (Wett & Rauch 2003; Guisasola *et al.* 2007; Al-Omari *et al.* 2015; Seuntjens *et al.* 2018). From a fundamental standpoint, NOB is likely less limited by inorganic carbon than AOB, as the latter can up-regulate its anabolism mixotrophically using traces of organic matter (Bock 1976). However, the different behaviour of AOB and NOB under inorganic carbon depleted conditions is typically neglected in models. Moreover, mainstream process models typically assume that influent inorganic carbon content is sufficiently high, meaning hardly any limitation (Sin *et al.* 2008). Bicarbonate as a state variable should be explicitly used to quantify inorganic carbon limitation rather than to simply indicate pH changes. In this case, sigmoidal kinetics are recommended. Furthermore, a higher inorganic carbon limitation for AOB than NOB is essential for modelling nitrogen conversions in trickling filters following direct anaerobic sewage treatment (Bressani-Ribeiro *et al.* 2021).

4.4. Denitrification intermediates (electron acceptors) and electron donors

Denitrification refers to the 4-steps reduction of NO_3^--N to N_2 via NO_2^--N, NO, and N_2O. While formerly considered in denitrification models as a 1-step process, heterotrophic denitrifiers and some phosphate accumulating organisms possess a highly modular microbiome with a diverse distribution of the nar, nir, nor, and nosZ genes (Ekama & Wentzel 1999; Graf *et al.* 2014). External carbon sources such as acetate, ethanol or methanol are being added to enhance denitrification rates (Mokhayeri *et al.* 2009). The metabolic pathways to oxidize each carbon source are different: the tricarboxylic acid cycle for acetate, specific enzymes to convert ethanol to acetate, and two other pathways requiring specific enzymes for methanol degradation (Ribera-Guardia *et al.* 2014). Hence, the chemical composition of the electron donor pool will shape the microbial community yielding significantly different denitrification rates and yields (Hallin *et al.* 2006; Lu *et al.* 2014). In an enriched denitrifying community the individual addition of excess acetate, ethanol, and methanol showed distinctive NO_3^--N reduction rates, and more importantly, different accumulation of the intermediates NO_2^--N and N_2O (Ribera-Guardia *et al.* 2014). Consequently, the oxidation rate of each carbon source can be the limiting step to the overall denitrification rate even at non-limiting organic carbon concentrations (Gao *et al.* 2020).

The distinct reduction and accumulation of denitrification intermediates depending on the carbon source is crucial especially for the PdNA systems since the partial denitrification rate is the key factor for the process performance. Two-step denitrification models fit the purpose of modelling short-cut N removal processes, but the ASMN extended to 4-steps to incorporate NO and N_2O reduction based on Von Schulthess *et al.* (1995). The current structures based on ASMN assume that carbon oxidation supplies all the electrons necessary for the four denitrification steps, leading to include individual maximum denitrification rates, substrate affinity and inhibition constants (e.g. NO, DO). Hence, the complexity of 4-step denitrification models increases significantly while datasets for the intermediates NO and N_2O remain scarce. Thus assumptions need to be made on substrate affinity kinetics for NO and N_2O reduction (Hiatt & Grady 2008). The specific maximum reduction rates for NO_3^--N and NO_2^--N vary substantially depending on the carbon source, and are typically modified during model calibrations to fit denitrification rates (Q. Wang *et al.* 2016; Domingo-Félez & Smets 2020). However, the approach of additive kinetics in ASMN fails to properly describe the electron competition (Pan *et al.* 2015). To overcome this challenge, the ASM-ICE (indirect coupling of electrons) was introduced to explicitly calculate the concentration of internal electron carriers of a methanol-enriched denitrifying community uncoupling the denitrification and carbon oxidation processes over one branch at a cost of higher model complexity (Pan *et al.* 2013). Based on Almeida *et al.* (1997) the ASM−EC (electron competition) calculates denitrification rates and carbon oxidation as an analogy to current intensity flowing through a parallel set of resistors in electric circuits. The ASM−EC was validated with data from batch experiments with four different carbon sources including acetate, ethanol, methanol, and their ternary mixture with fewer parameters than the ASM-ICE (Domingo-Félez & Smets 2020).

4.5. Inclusion of N2O emission

N_2O models have been developed in biological nitrogen removal systems and reached maturity to facilitate process optimization of conventional N removal systems (Section 2.4). However, they have not been widely applied in modelling short-cut processes. The lack of datasets is one of the major challenges to applying N_2O models and particularly the datasets from large-scale demonstrations. For that reason, monitoring the N_2O emissions is essential in pilot and full-scale applications to improve the applicability of N_2O models in mainstream short-cut processes. Hybrid models can also be adopted for short-term laboratory or pilot-scale studies with data scarcity where mechanistic models can be used to simulate the biological process and the data generated can be used in a data-driven model that acts as input to a N_2O prediction algorithm (Mehrani *et al.* 2022).

N_2O modelling in mainstream short-cut N removal systems shares similar challenges as in conventional nitrification and 2-step denitrification systems. For example, since N_2O is an intermediate in the nitrogen transformation pathways, N_2O models require a laborious calibration process under varying operational conditions (Hwangbo *et al.* 2021). Also, microbial communities involved in N_2O production are more complex than conventional systems (e.g. NO is an intermediate of the anammox metabolism and the precursor of N_2O for AOB and heterotrophic denitrifiers). This leads to overparameterized N_2O process models for the description of biological pathways. For example, the aforementioned preferential uptake of carbon sources leads to different patterns of NO_2^- accumulation during denitrification (Ribera-Guardia *et al.* 2014; Zhao *et al.* 2018) and impacts N_2O model calibration (Domingo-Félez & Smets 2020). Thus, while it is important to include the N_2O models for modelling of short-cut processes, model complexity and ease-of-use should be prioritized based on the modelling goals.

5. CONCLUSION AND PERSPECTIVES

A review was presented on models for short-cut N removal processes, in view of mainstream applications, considering the current practice, limitations, future needs and perspectives. Mathematical models are under development to overcome implementation challenges and to deal with the complexity of mainstream deammonification. The pilot and full-scale applications reported provide invaluable information for models development while modelling efforts can accelerate the success of practical applications.

- NOB out-selection is still a major issue for mainstream application. Its dynamics can be captured in individual models but there are still limitations such as the model parameter values that may vary depending on the process configuration and operating conditions. Thus, they may not be transferable to another system.
- Recent modelling efforts mostly deal with the application of mainstream deammonification in full-scale and there appears to be less focus on nitrite-shunt.
- To correctly model the N mass balance and anammox process performance, models should consider incorporation of heterotrophic denitrifiers and anammox bacteria and the yield for the overall anammox metabolic reaction based on ammonium take up.
- While steady-state simulations have been used to model the overall feasibility and performance of short-cut N removal systems, dynamic simulations are especially needed to address issues with capacity evaluations, control strategies for sustained performance, and assessing the risk for meeting effluent limits.
- The application of robust sensitivity analysis is urgently needed and more rigorous calibration methodologies should be adopted to allow the comparison between model structures. The simulated model outputs would become much more reproducible and transferable.
- Modelling PdNA requires accurate modelling of partial denitrification and the ability to address microbial competition among the electron acceptors for various carbon sources. The existing PdNA models were only tested thus far to simulate partial denitrification with simple carbon sources. Other electron donors need to be better characterized and evaluated

using the models to simulate PdNA systems that rely on complex influent COD, fermentation products or internally stored carbon.

- Competition over different substrates and estimation of half-saturation constants should be handled carefully by considering system properties such as mixing, diffusion processes and substrate limitations such as by nitrite and inorganic carbon availability to properly model the substrate competition and out-selection. Hybrid models can be adopted by adding a data-driven component to the well-described mechanistic model structure to model substrate competition properly and compensate the lack of mechanistic description of subprocesses influencing the kinetics of short-cut processes.
- The inclusion of interspecies competition in models by implementing different genera of the same functional group (with different kinetic properties) may be warranted in cases where it affects the macroscopic reactor performance in order to predict a more reliable NOB control performance, and thus find conditions for a more successful process start-up and stable mainstream short-cut process.
- The denitrification intermediate steps are crucial especially for partial denitrification efficiency and N_2O emissions. In model development, 2-step denitrification fits the purpose of modelling short-cut process efficiency while more detailed 4-step models should be adopted if the modelling goal is to predict both process efficiency and N_2O emissions.

ACKNOWLEDGEMENTS

This work is an outcome of discussions at the 'Mainstream Short-cut Nitrogen Removal Modelling: From research to full-scale implementation, do we have what we need?' Workshop of the 7th IWA/WEF Water Resource Recovery Modelling Seminar. It is a collective effort by the practitioners and academics of the wastewater industry who are keen to implement the deammonification process sustainably on mainstream treatment.

DECLARATION OF COMPETING INTEREST

The authors declare that they have no known competing financial interests or personal relationships that could influence the work reported in this paper.

DATA AVAILABILITY STATEMENT

All relevant data are included in the paper or its Supplementary Information.

REFERENCES

Agrawal, S., Seuntjens, D., Cocker, P. D., Lackner, S. & Vlaeminck, S. E. 2018 Success of mainstream partial nitritation/anammox demands integration of engineering, microbiome and modeling insights. *Current Opinion in Biotechnology* **50**, 214–221. doi:10.1016/j.copbio.2018.01.013.

Al-Hazmi, H. E., Yin, Z., Grubba, D., Majtacz, J. B. & Makinia, J. 2022 Comparison of the efficiency of deammonification under different DO concentrations in a laboratory-scale sequencing batch reactor. *Water* **14** (3), 368. doi:10.3390/w14030368.

Ali, P., Zalivina, N., Le, T., Riffat, R., Ergas, S., Wett, B., Murthy, S., Al-Omari, A., deBarbadillo, C., Bott, C. & De Clippeleir, H. 2020 Primary sludge fermentate as carbon source for mainstream partial denitrification – anammox (PdNA). *Water Environment Research* **93** (7), 1044–1059. doi:10.1002/wer.1492.

Almeida, J. S., Reis, M. A. M. & Carrondo, M. J. T. 1997 A unifying kinetic model of denitrification. *Journal of Theoretical Biology* **186** (2), 241–249. doi:10.1006/jtbi.1996.0352.

Al-Omari, A., Wett, B., Nopens, I., De Clippeleir, D., Han, M., Regmi, P., Bott, C. & Murthy, S. 2015 Model-based evaluation of mechanisms and benefits of mainstream shortcut nitrogen removal processes. *Water Science and Technology* **71** (6), 840–847. doi:10.2166/wst.2015.022.

Al-Omari, A., De Clippeleir, H., Ladipo-Obasa, M., Klaus, S., Bott, C. B., McCullough, K., Fofana, R., Wadhawan, T., Murthy, S., Fevig, S., Jimenez, J., Wett, B. & Nopens, I. 2021 Modelling partial heterotrophic denitrification in mainstream nitrogen removal processes – model development and evaluation. In: *Proceedings of 7th IWA Water Resource Recovery Modelling Seminar*. International Water Association, pp. 67–71.

Amaral, A., Gillot, S., Garrido-Baserba, M., Filali, A., Karpinska, A. M., Plósz, B. G., De Groot, C., Bellandi, G., Nopens, I., Takács, I., Lizarralde, I., Jimenez, J. A., Fiat, J., Rieger, L., Arnell, M., Andersen, M., Jeppsson, U., Rehman, U., Fayolle, Y., Amerlinck, Y. & Rosso, R. 2019 Modelling gas–liquid mass transfer in wastewater treatment: when current knowledge needs to encounter engineering practice and vice versa. *Water Science and Technology* **80** (4), 607–619. doi:10.2166/wst.2019.253.

Arnaldos, M., Amerlinck, Y., Rehman, U., Maere, T., Van Hoey, S., Naessens, W. & Nopens, I. 2015 From the affinity constant to the half-saturation index: understanding conventional modeling concepts in novel wastewater treatment processes. *Water Research* **70**, 458–470. doi:10.1016/j.watres.2014.11.046.

Arnell, M., Rahmberg, M., Oliveira, F. & Jeppsson, U. 2017 Multi-objective performance assessment of wastewater treatment plants combining plant-wide process models and life cycle assessment. *Journal of Water and Climate Change* **8** (4), 715–729. doi:10.2166/wcc.2017.179.

Baek, S. H. & Kim, H. J. 2013 Mathematical model for simultaneous nitrification and denitrification (SND) in membrane bioreactor (MBR) under low dissolved oxygen (DO) concentrations. *Biotechnology and Bioprocess Engineering* **18**, 104–110. doi:10.1007/s12257-011-0419-6.

Baeten, J. E., Batstone, D. J., Schraa, O. J., van Loosdrecht, M. C. M. & Volcke, E. I. P. 2019 Modelling anaerobic, aerobic and partial nitritation-anammox granular sludge reactors – a review. *Water Research* **149**, 322–341. doi:10.1016/j.watres.2018.11.026.

Biesterfeld, S., Farmer, G., Russel, P. & Figueroa, L. 2003 Effect of alkalinity type and concentration on nitrifying biofilm activity. *Water Environment Research* **75** (3), 196–204. Available from: https://www.jstor.org/stable/25045684

Blackburne, R., Yuan, Z. & Keller, J. 2008 Demonstration of nitrogen removal via nitrite in a sequencing batch reactor treating domestic wastewater. *Water Research* **42**, 2166–2176. doi:10.1016/j.watres.2007.11.029.

Bock, E. 1976 Growth of *Nitrobacter* in the presence of organic matter. *Archives of Microbiology* **108** (3), 305–312. doi: 10.1007/BF00454857.

Boltz, J. P., Morgenroth, E., Brockmann, D., Bott, C., Gellner, W. J. & Vanrolleghem, P. A. 2011 Systematic evaluation of biofilm models for engineering practice: components and critical assumptions. *Water Science and Technology* **64** (4), 930–944. doi:10.2166/wst.2011.709.

Bougard, D., Bernet, N., Dabert, P., Delgenes, J. P. & Steyer, J. P. 2006 Influence of closed loop control on microbial diversity in a nitrification process. *Water Science and Technology* **53** (4–5), 85–93. doi:10.2166/wst.2006.113.

Bournazou, M. C., Hooshiar, K., Arellano-Garcia, H., Wozny, G. & Lyberatos, G. 2013 Model based optimization of the intermittent aeration profile for SBRs under partial nitrification. *Water Research* **47** (10), 3399–3410. doi:10.1016/j.watres.2013.03.044.

Bozileva, E., Khiewwijit, R., Temmink, H., Rijnaarts, H. & Keesman, K. 2017 Exploring the feasibility of a novel municipal wastewater treatment system via dynamic plant-wide simulation. In: *Frontiers in Wastewater Treatment and Modelling. FICWTM 2017. Lecture Notes in Civil Engineering*, vol. 4 (Mannina, G., ed.). Springer, Cham, pp. 575–582. doi:10.1007/978-3-319-58421-8_90.

Bressani-Ribeiro, T., Almeida, P. G. S., Chernicharo, C. A. L. & Volcke, E. I. P. 2021 Inorganic carbon limitation during nitrogen conversions in sponge-bed trickling filters for mainstream treatment of anaerobic effluent. *Water Research* **201**, 117337. doi:10.1016/j.watres.2021.117337.

Campolong, C., Klaus, S., Ferguson, L., Wilson, C., Wett, B., Murthy, S. & Bott, C. B. 2018 Optimizing carbon addition to a polishing partial denitrification/anammox MBBR using online control. In: *Proceedings of the Water Environment Federation*. Water Environment Federation, pp. 164–168. doi:10.2175/193864718824940565.

Cao, S., Wang, S., Peng, Y., Wu, C., Du, R., Gong, L. & Ma, B. 2013 Achieving partial denitrification with sludge fermentation liquid as carbon source: the effect of seeding sludge. *Bioresource Technology* **149**, 570–574. doi:10.1016/j.biortech.2013.09.072.

Cao, S., Oehmen, A. & Zhou, Y. 2019 Denitrifiers in mainstream anammox processes: competitors or supporters? *Environmental Science and Technology* **53** (9), 11063–11065. doi:10.1021/acs.est.9b05013.

Cao, Y., van Loosdrecht, M. C. M. & Daigger, G. T. 2017 Mainstream partial nitration-anammox in municipal wastewater treatment: status, bottlenecks, and further studies. *Applied Microbiology and Biotechnology* **101** (4), 1365–1383. doi:10.1007/s00253-016-8058-7.

Castro-Barros, C. M., Rodríguez-Caballero, A., Volcke, E. I. P. & Pijuan, M. 2016 Effect of nitrite on the N2O and NO production on the nitrification of low-strength ammonium wastewater. *Chemical Engineering Journal* **287**, 269–276. https://doi.org/10.1016/j.cej.2015.10.121.

Chen, H., Tu, Z., Wu, S., Yu, G., Du, C., Wang, H., Yang, E., Zhou, L., Deng, B., Wang, D. & Li, H. 2021 Recent advances in partial denitrification-anaerobic ammonium oxidation process for mainstream municipal wastewater treatment. *Chemosphere* **278**, 30436. doi:10.1016/j.chemosphere.2021.130436.

Chen, X., Mielczarek, A. T., Habicht, K., Andersen, M. H., Thornberg, D. & Sin, G. 2019 Assessment of full-scale N$_2$O emission characteristics and testing of control concepts in an activated sludge wastewater treatment plant with alternating aerobic and anoxic phases. *Environmental Science and Technology* **53** (21), 12485–12494. doi:10.1021/acs.est.9b04889.

Cosenza, A., Mannina, G., Vanrolleghem, P. A. & Neumann, M. B. 2013 Global sensitivity analysis in wastewater applications: a comprehensive comparison of different methods. *Environmental Modelling & Software* **49**, 40–52. https://doi.org/10.1016/j.envsoft.2013.07.009.

Cui, F., Park, S., Mo, K., Lee, W., Lee, H. & Kim, M. 2017 Experimentation and mathematical models for partial nitrification in aerobic granular sludge process. *KSCE Journal of Civil Engineering* **21**, 127–133. doi:10.1007/s12205-016-0506-5.

Daigger, G. T. 2014 Oxygen and carbon requirements for biological nitrogen removal processes accomplishing nitrification, nitritation, and anammox. *Water Environment Research* **86** (3), 204–209. doi:10.2175/106143013(13807328849459.

De Clippeleir, H., Vlaeminck, S. E., De Wilde, F., Daeninck, K., Mosquera, M., Boeckx, P., Verstraete, W. & Boon, N. 2013 One-stage partial nitritation/anammox at 15 °C on pretreated sewage: feasibility demonstration at lab-scale. *Applied Microbiology and Biotechnology* **97** (23), 10199–10210. doi:10.1007/s00253-013-4744-x.

De Kreuk, M. K., Picioreanu, C., Hosseini, M., Xavier, J. B. & van Loosdrecht, M. C. M. 2007 Kinetic model of a granular sludge SBR: influences on nutrient removal. *Biotechnology and Bioengineering* **97** (4), 801–815. doi:10.1002/bit.21196.

De Pauw, D. J. & Vanrolleghem, P. A. 2006a Practical aspects of sensitivity function approximation for dynamic models. *Mathematical and Computer Modelling of Dynamical Systems* **12** (5), 395–414. doi:10.1080/13873950600723301.

De Pauw, D. J. & Vanrolleghem, P. A. 2006b Designing and performing experiments for model calibration using an automated iterative procedure. *Water Science and Technology* **53** (1), 117–127. https://doi.org/10.2166/wst.2006.014.

Domingo-Félez, C. & Smets, B. F. 2016 A consilience model to describe N2O production during biological N removal. *Environmental Science: Water Research & Technology* **2** (6), 923–930. doi:10.1039/C6EW00179C.

Domingo-Félez, C. & Smets, B. F. 2020 Modeling denitrification as an electric circuit accurately captures electron competition between individual reductive steps: the activated sludge model–electron competition model. *Environmental Science & Technology* **54** (12), 7330–7338. doi:10.1021/acs.est.0c01095.

Domingo-Félez, C., Mutlu, A. G., Jensen, M. M. & Smets, B. F. 2014 Aeration strategies to mitigate nitrous oxide emissions from single-stage nitritation/Anammox reactors. *Environmental Science and Technology* **48** (15), 8679–8687. doi:10.1021/es501819n.

Drewnowski, J., Shourjeh, M. S., Kowal, P. & Cel, W. 2020 Modelling AOB-NOB competition in shortcut nitrification compared with conventional nitrification-denitrification process. In: *Journal of Physics: Conference Series, Volume 1736, V International Conference of Computational Methods in Engineering Science – CMES'20*, Lublin, Poland – Lviv, Ukraine.

Du, R., Cao, S., Li, B., Niu, M., Wang, S. & Peng, Y. 2017 Performance and microbial community analysis of a novel DEAMOX based on partial-denitrification and anammox treating ammonia and nitrate wastewaters. *Water Research* **108**, 46–65. doi:10.1016/j.watres.2016.10.051.

Du, R., Peng, Y., Ji, J., Shi, L., Gao, R. & Li, X. 2019 Partial denitrification providing nitrite: opportunities of extending application for anammox. *Environment International* **131**, 105001. doi:10.1016/j.envint.2019.105001.

Duan, H., Ye, L., Erler, D., Ni, B. J. & Yuan, Z. 2017 Quantifying nitrous oxide production pathways in wastewater treatment systems using isotope technology – a critical review. *Water Research* **122**, 96–113. doi:10.1016/j.watres.2017.05.054.

Duan, H., Ye, L., Lu, X. & Yuan, Z. 2019a Overcoming nitrite oxidizing bacteria adaptation through alternating sludge treatment with free nitrous acid and free ammonia. *Environmental Science and Technology* **53** (4), 1937–1946. doi:10.1021/acs.est.8b06148.

Duan, H., Ye, L., Wang, Q., Zheng, M., Lu, X., Wang, Z. & Yuan, Z. 2019b Nitrite oxidizing bacteria (NOB) contained in influent deteriorate mainstream NOB suppression by sidestream inactivation. *Water Research* **162**, 331–338. doi:10.1016/j.watres.2019.07.002.

Duan, H., van den Akker, B., Thwaites, B. J., Peng, L., Herman, C., Pan, Y., Ni, B.-J., Watt, S., Yuan, Z. & Ye, L. 2020 Mitigating nitrous oxide emissions at a full-scale wastewater treatment plant. *Water Research* **185**, 116196. doi:10.1016/j.watres.2020.116196.

Duan, H., Zhao, Y., Koch, K., Wells, G. F., Zheng, M., Yuan, Z. & Ye, L. 2021 Insights into nitrous oxide mitigation strategies in wastewater treatment and challenges for wider implementation. *Environmental Science and Technology* **55** (11), 7208–7224. doi:10.1021/acs.est.1c00840.

Dytczak, M. A., Londry, K. L. & Oleszkiewicz, J. A. 2008 Activated sludge operational regime has significant impact on the type of nitrifying community and its nitrification rates. *Water Research* **42** (8–9), 2320–2328. doi:10.1016/j.watres.2007.12.018.

Edenhofer, O., Pichs-Madruga, R., Sokona, Y., Minx, J. C., Farahani, E., Kadner, S., Seyboth, K., Adler, A., Baum, I., Brunner, S., Eickemeier, P., Kriemann, B., Savolainen, J., Schlömer, S., von Stechow, C. & Zwickel, T. 2014 IPCC, 2014: Climate Change 2014: Mitigation of Climate Change. In: *Contribution of Working Group III to the Fifth Assessment Report of the Intergovernmental Panel on Climate Change*. Cambridge University Press, Cambridge, United Kingdom and New York, USA. https://www.ipcc.ch/report/ar5/wg3/.

Ekama, G. A. & Wentzel, C. 1999 Denitrification kinetics in biological N and P removal activated sludge systems treating municipal wastewaters. *Water Science and Technology* **39** (6), 69–77. doi:10.1016/S0273-1223(99)00124-9.

Ganigué, R., Gabarró, J., Sànchez-Melsió, A., Ruscalleda, M., López, H., Vila, X., Colprim, J. & Dolors Balaguer, M. 2009 Long-term operation of a partial nitritation pilot plant treating leachate with extremely high ammonium concentration prior to an anammox process. *Bioresource Technology* **100** (23), 5624–5632. doi:10.1016/j.biortech.2009.06.023.

Gao, D. & Xiang, T. 2021 Deammonification process in municipal wastewater treatment: challenges and perspectives. *Bioresource Technology* **320**, 124420. doi:10.1016/j.biortech.2020.124420.

Gao, H., Zhao, X., Zhou, L., Sabba, F. & Wells, G. F. 2020 Differential kinetics of nitrogen oxides reduction leads to elevated nitrous oxide production by a nitrite fed granular denitrifying EBPR bioreactor. *Environmental Science: Water Research & Technology* **6** (4), 1028–1043. doi:10.1039/C9EW00881 K.

Gernaey, K., van Loosdrecht, M. C. M., Lind, M. & Jørgensen, S. B. 2004 Activated sludge wastewater treatment plant modelling and simulation: state of the art. *Environmental Modelling & Software* **19** (9), 763–783. doi:10.1016/j.envsoft.2003.03.005.

Gilbert, E. M., Agrawal, S., Brunner, F., Schwartz, T., Horn, H. & Lackner, S. 2014 Response of different *Nitrospira* species to anoxic periods. *Environmental Science and Technology* **48** (5), 2934–2941. doi:10.1021/es404992 g.

Gilbert, E. M., Agrawal, S., Schwartz, T., Horn, H. & Lackner, S. 2015 Comparing different reactor configurations for partial nitritation/anammox at low temperatures. *Water Research* **81**, 92–100. doi:10.1016/j.watres.2015.05.022.

Graf, D. R., Jones, C. M. & Hallin, S. 2014 Intergenomic comparisons highlight modularity of the denitrification pathway and underpin the importance of community structure for N2O emissions. *PLoS One* **9** (12), e114118. eCollection 2014. doi:10.1371/journal.pone.0114118.

Guisasola, A., Petzet, S., Baeza, J., Carrera, J. & Lafuente, J. 2007 Inorganic carbon limitations on nitrification: experimental assessment and modelling. *Water Research* **41** (2), 277–286. doi:10.1016/j.watres.2006.10.030.

Guo, L. & Vanrolleghem, P. 2014 Calibration and validation of an activated sludge model for greenhouse gases no. 1 (ASMG1): prediction of temperature-dependent N2O emission dynamics. *Bioprocess and Biosystem Engineering* **37** (2), 151–163. doi:10.1007/s00449-013-0978-3.

Gustavsson, D. J., Suarez, C., Wilén, B. M., Hermansson, M. & Persson, F. 2020 Long-term stability of partial nitritation-anammox for treatment of municipal wastewater in a moving bed biofilm reactor pilot system. *Science of Total Environment* **714**, 136342. doi:10.1016/j.scitotenv.2019.136342.

Hallin, S., Throbäck, I. N., Dicksved, J. & Pell, M. 2006 Metabolic profiles and genetic diversity of denitrifying communities in activated sludge after addition of methanol or ethanol. *Applied and Environmental Microbiology* **72** (8), 5445–5452. doi:10.1128/AEM.00809-06.

Han, M., Clippeleir, H. D., Al-Omari, A., Wett, B., Vlaeminck, S., Bott, C. & Murthy, S. 2016 Impact of carbon to nitrogen ratio and aeration regime on mainstream deammonification. *Water Science and Technology* **74** (2), 375–384. doi:10.2166/wst.2016.202.

Hao, X. D. & van Loosdrecht, M. C. M. 2004 Model-based evaluation of COD influence on a partial nitrification-Anammox biofilm (CANON) process. *Water Science and Technology* **49** (11–12), 83–90. https://doi.org/10.2166/wst.2004.0810.

Hao, X., Heijnen, J. J. & van Loosdrecht, M. C. M. 2002 Sensitivity analysis of a biofilm model describing a one-stage completely autotrophic nitrogen removal (CANON) process. *Biotechnology and Bioengineering* **77** (3), 266–277. doi:10.1002/bit.10105.

Hellinga, C., van Loosdrecht, M. C. M. & Heijnen, J. J. 1999 Model based design of a novel process for nitrogen removal from concentrated flows. *Mathematical and Computer Modelling of Dynamical Systems* **5** (4), 351–371. doi:10.1076/mcmd.5.4.351.3678.

Henze, M., Gujer, W., Mino, T. & van Loosdrecht, M. C. M. 2006 *Activated Sludge Models ASM1, ASM2, ASM2d and ASM3*. IWA Publishing, Cornwall, UK. doi:10.2166/9781780402369.

Hiatt, W. & Grady, C. 2008 An updated process model for carbon oxidation, nitrification, and denitrification. *Water Environment Research* **80** (11), 2145–2156. doi:10.2175/106143008(304776.

Hoekstra, M., Geilvoet, S., Hendrickx, T., Erp Taalman Kip, C. V., Kleerebezem, R. & van Loosdrecht, M. C. M. 2018 Towards mainstream anammox: lessons learned from pilot-scale research at WWTP Dokhaven. *Environmental Technology* **40** (13), 1721–1733. doi:10.1080/09593330.2018.1470204.

Huang, X., Mi, W., Hong, N., Ito, H. & Kawagoshi, Y. 2020 Efficient transition from partial nitritation to partial nitritation/Anammox in a membrane bioreactor with activated sludge as the sole seed source. *Chemosphere* **253**, 126719. doi:10.1016/j.chemosphere.2020.126719.

Hubaux, N., Wells, G. & Morgenroth, E. 2015 Impact of coexistence of flocs and biofilm on performance of combined nitritation-anammox granular sludge reactors. *Water Research* **68**, 127–139. doi:10.1016/j.watres.2014.09.036.

Hwangbo, S., Al, R., Chen, X. & Sin, G. 2021 Integrated model for understanding N2O emissions from wastewater treatment plants: a deep learning approach. *Environmental Science and Technology* **55** (3), 2143–2151. doi:10.1021/acs.est.0c05231.

Jauffur, S., Isazadeh, S. & Frigon, D. 2014 Should activated sludge models consider influent seeding of nitrifiers? Field characterization of nitrifying bacteria. *Water Science and Technology* **70** (9), 1526–1532. doi:10.2166/wst.2014.407.

Jetten, M. S., Horn, S. J. & van Loosdrecht, M. C. M. 1997 Towards a more sustainable municipal wastewater treatment system. *Water Science and Technology* **35** (9), 171–180. doi:10.1016/S0273-1223(97)00195-9.

Ji, J., Peng, Y., Wang, B. & Wang, S. 2017 Achievement of high nitrite accumulation via endogenous partial denitrification (EPD). *Bioresource Technology* **224**, 140–146. doi:10.1016/j.biortech.2016.11.070.

Ji, J., Peng, Y., Mai, W., He, J., Wang, B., Li, X. & Zhang, Q. 2018 Achieving advanced nitrogen removal from low C/N wastewater by combining endogenous partial denitrification with anammox in mainstream treatment. *Bioresource Technology* **270**, 570–579. doi:10.1016/j.biortech.2018.08.124.

Ji, J., Peng, Y., Wang, B., Li, X. & Zhang, Q. 2020 Synergistic partial-denitrification, anammox, and in-situ fermentation (SPDAF) process for advanced nitrogen removal from domestic and nitrate-containing wastewater. *Environmental Science & Technology* **54** (6), 3702–3713. doi:10.1021/acs.est.9b07928.

Ji, Z. & Chen, Y. 2010 Using sludge fermentation liquid to improve wastewater short-cut nitrification-denitrification and denitrifying phosphorus removal via nitrite. *Environmental Science and Technology* **44** (23), 8957–8963. doi:10.1021/es102547n.

Jia, M., Castro-Barros, C. M., Winkler, M. K. H. & Volcke, E. I. P. 2018 Effect of organic matter on the performance and N2O emission of a granular sludge anammox reactor. *Environmental Science: Water Research and Technology* **4**, 1035–1046. doi:10.1039/C8EW00125A.

Jia, M., Solon, K., Vandeplassche, D., Venugopal, H. & Volcke, E. I. P. 2020 Model-based evaluation of an integrated high-rate activated sludge and mainstream anammox system. *Chemical Engineering Journal* **382**, 122878. doi:10.1016/j.cej.2019.122878.

Jimenez, J., Wise, G., Regmi, P., Burger, G., Conidi, D., Du, W. & Dold, P. 2020 Nitrite-shunt and biological phosphorus removal at low dissolved oxygen in a full-scale high-rate system at warm temperatures. *Water Environment Research* **92** (8), 1111–1122. doi:10.1002/wer.1304.

Kartal, B., Kuenen, J. G. & Loosdrecht, M. C. M. 2010 Sewage treatment with anammox. *Science* **328** (5979), 702–703. doi:10.1126/science.1185941.

Klaus, S. A. 2019 *Intensification of Biological Nutrient Removal Processes*. PhD Dissertation, Virginia Polytechnic Institute and State University, Virginia, USA.

Klaus, S., Parsons, M. & Bott, C. 2020 Mainstream partial denitrification/anammox: results from operation in a full-scale deep-bed filter. In: *Proceedings of IWA Nutrient Removal and Recovery Conference 2020*, Helsinki, Finland.

Lackner, S. & Agrawal, S. 2015 Process fundamentals – microbiology, stoichiometry, kinetics, and inhibition. In: *Shortcut Nitrogen Removal – Nitrite Shunt and Deammonification*. Water Environment Federation. https://www.accesswater.org/?id=-323517.

Lackner, S. & Smets, B. F. 2012 Effect of the kinetics of ammonium and nitrite oxidation on nitritation success or failure for different biofilm reactor geometries. *Biochemical Engineering Journal* **69**, 123–129. doi:10.1016/j.bej.2012.09.006.

Lackner, S., Gilbert, E., Vlaeminck, S., Joss, A., Horn, H. & van Loosdrecht, M. C. M. 2014 Full-scale partial nitritation/anammox experiences – an application survey. *Water Research* **55**, 292–303. doi:10.1016/j.watres.2014.02.032.

Laureni, M., Weissbrodt, D., Villez, K., Robin, O., de Jonge, N., Rosenthal, A., Wells, G., Nielsen, J. N., Morgenroth, E. & Joss, A. 2019 Biomass segregation between biofilm and flocs improves the control of nitrite-oxidizing bacteria in mainstream partial nitritation and anammox processes. *Water Research* **154**, 104–116. doi:10.1016/j.watres.2018.12.051.

Le, T., Peng, B., Su, C., Massoudieh, A., Torrents, A., Al-Omari, A., Murthy, S., Wett, B., Chandran, K., deBarbadillo, C., Bott, C. & De Clippeleir, H. 2019a Nitrate residual as a key parameter to efficiently control partial denitrification coupling with anammox. *Water Environment Research* **91**, 1455–1465. doi:10.1002/wer.1140.

Le, T., Peng, B., Su, C., Massoudieh, A., Torrents, A., Al-Omari, A., Murthy, S., Wett, B., Chandran, K., deBarbadillo, C., Bott, C. & De Clippeleir, H. 2019b Impact of carbon source and COD/N on the concurrent operation of partial denitrification and anammox. *Water Environment Research* **91** (3), 185–197.

Le, T., Fofana, R., Massoudieh, A., Al-Omari, A., Murthy, S., Wett, B., Chandran, K., deBarbadillo, C., Bott, C. & De Clippeleir, H. 2019c A novel online dynamic control coupling AvN and PdN-AnAOB to achieve stringent effluent limits in mainstream during wet weather. In: *Proceedings of the Water Environment Federation, WEFTEC 2019*. Water Environment Federation, pp. 1562–1573.

Leal, C. D., Pereira, A. D., Nunes, F. T., Ferreira, L. O., Coelho, A. C., Bicalho, S. K., Abreu Mac Conell, E. F., Bressani Ribeiro, T., de Lemos Chernicharo, C. A. & de Araújo, J. C. 2016 Anammox for nitrogen removal from anaerobically pre-treated municipal wastewater: effect of COD/N ratios on process performance and bacterial community structure. *Bioresource Technology* **211**, 257–266. doi:10.1016/j.biortech.2016.03.107.

Ledergerber, J. M., Maruéjouls, T. & Vanrolleghem, P. A. 2019 Optimal experimental design for calibration of a new sewer water quality model. *Journal of Hydrology* **574**, 1020–1028. doi:10.1016/j.jhydrol.2019.05.004.

Li, J., Peng, Y., Zhang, L., Liu, J., Wang, X., Gao, R., Pang, L. & Zhou, Y. 2019 Quantify the contribution of anammox for enhanced nitrogen removal through metagenomic analysis and mass balance in an anoxic moving bed biofilm reactor. *Water Research* **160**, 178–187. doi:10.1016/j.watres.2019.05.070.

Li, L., Ling, Y., Wang, H., Chu, Z., Yan, G., Li, Z. & Wu, T. 2020 N_2O emission in partial nitritation-anammox process. *Chinese Chemical Letters* **31** (1), 28–38. doi:10.1016/j.cclet.2019.06.035.

Liu, G. & Wang, J. 2013 Long-Term low DO enriches and shifts nitrifier community in activated sludge. *Environmental Science & Technology* **47** (10), 5109–5117. doi:10.1021/es304647y.

Liu, T., Ma, B., Chen, X., Ni, B.-J., Peng, Y. & Guo, J. 2017 Evaluation of mainstream nitrogen removal by simultaneous partial nitrification, anammox and denitrification (SNAD) process in a granule-based reactor. *Chemical Engineering Journal* **327**, 973–981. doi:10.1016/j.cej.2017.06.173.

Liu, X., Kim, M., Nakhla, G., Andalib, M. & Fang, Y. 2020 Partial nitrification-reactor configurations, and operational conditions: performance analysis. *Journal of Environmental Chemical Engineering* **8** (4), 103984. doi:10.1016/j.jece.2020.103984.

Liu, W., Hao, S., Ma, B., Zhang, S. & Li, J. 2022 In-situ fermentation coupling with partial-denitrification/anammox process for enhanced nitrogen removal in an integrated three-stage anoxic/oxic (A/O) biofilm reactor treating low COD/N real wastewater. *Bioresource Technology* **344**, 126267. doi:10.1016/j.biortech.2021.126267.

Lotti, T., Kleerebezem, R., Hu, Z., Kartal, B., Kreuk, M. K., van Erp Taalman Kip, C., Kruit, J., Hendrickx, T. L. G. & van Loosdrecht, M. C. M. 2015 Pilot-scale evaluation of anammox-based mainstream nitrogen removal from municipal wastewater. *Environmental Technology* **36** (9), 1167–1177. doi:10.1080/09593330.2014.982722.

Lu, H., Chandran, K. & Stensel, D. 2014 Microbial ecology of denitrification in biological wastewater treatment. *Water Research* **64**, 237–254. doi:10.1016/j.watres.2014.06.042.

Lu, W., Bin, M., Wang, Q., Wei, Y. & Su, Z. 2021a Feasibility of achieving advanced nitrogen removal via endogenous denitratation/anammox. *Bioresource Technology* **325**, 124666. doi:10.1016/j.biortech.2021.124666.

Lu, W., Zhang, Y., Wang, Q., Wei, Y., Bu, Y. & Ma, B. 2021b Achieving advanced nitrogen removal in a novel partial denitrification/anammox-nitrifying (PDA-N) biofilter process treating low C/N ratio municipal wastewater. *Bioresource Technology* **340**, 125661. doi:10.1016/j.biortech.2021.125661.

Ma, B., Qian, W., Yuan, C., Yuan, Z. & Peng, Y. 2017 Achieving mainstream nitrogen removal through coupling anammox with denitratation. *Environmental Science and Technology* **51** (15), 8405–8413. doi:10.1021/acs.est.7b01866.

Ma, B., Xu, X., Wei, Y., Ge, C. & Peng, Y. 2020 Recent advances in controlling denitritation for achieving denitratation/anammox in mainstream wastewater treatment plants. *Bioresource Technology* **299**, 122697. doi:10.1016/j.biortech.2019.122697.

Ma, W.-J., Li, G.-F., Huang, B.-C. & Jin, R.-C. 2020 Advances and challenges of mainstream nitrogen removal from municipal wastewater with anammox-based processes. *Water Environment Research* **92** (11), 1899–1909. doi:10.1002/wer.1342.

Ma, Y., Domingo-Félez, C., Plósz, B. G. & Smets, B. F. 2017 Intermittent aeration suppresses nitrite-oxidizing bacteria in membrane-aerated biofilms: a model-based explanation. *Environmental Science and Technology* **51** (11), 6146–6155. doi:10.1021/acs.est.7b00463.

Ma, Y., Piscedda, A., Veras, A. D., Domingo-Félez, C. & Smets, B. F. 2021 Intermittent aeration to regulate microbial activities in membrane-aerated biofilm reactors: energy-efficient nitrogen removal and low nitrous oxide emission. *Chemical Engineering Journal* 133630. (In Press). doi:10.1016/j.cej.2021.133630.

Ma, Y., Sundar, S., Park, H. & Chandran, K. 2015 The effect of inorganic carbon on microbial interactions in a biofilm nitritation–anammox process. *Water Research* **70**, 246–254. doi:10.1016/j.watres.2014.12.006.

Mampaey, K. E., Beuckels, B., Kampschreur, M. J., Kleerebezem, R., van Loosdrecht, M. C. M. & Volcke, E. I. P. 2013 Modelling nitrous and nitric oxide emissions by autotrophic ammonia-oxidizing bacteria. *Environmental Technology* **34** (9–12), 1555–1566. doi:10.1080/09593330.2012.758666.

Mannina, G., Bella, G. D. & Viviani, G. 2010 Uncertainty assessment of a membrane bioreactor model using the GLUE methodology. *Biochemical Engineering Journal* **52** (2–3), 263–275. doi:10.1016/j.bej.2010.09.001.

Mannina, G., Ferreira Rebouças, T., Cosenza, A. & Chandran, K. 2019 A plant-wide wastewater treatment plant model for carbon and energy footprint: model application and scenario analysis. *Journal of Cleaner Production* **217**, 244–256. doi:10.1016/j.jclepro.2019.01.255.

Manser, R., Gujer, W. & Siegrist, H. 2005 Consequences of mass transfer effects on the kinetics of nitrifiers. *Water Research* **39** (19), 4633–4642. doi:10.1016/j.watres.2005.09.020.

McCullough, K., Klaus, S., Parsons, M. & Bott, C. 2021 The theoretical benefits of mainstream shortcut nitrogen removal revisited and validated by full-scale implementation of partial denitrification-anammox. In: *Proceedings of WEF Innovations in Process Engineering 2021 Virtual Event.* Water Environment Federation.

Mehrani, M. J., Bagherzadeh, F., Zheng, M., Kowal, P., Sobotka, D. & Makinia, J. 2022 Application of a hybrid mechanistic/machine learning model for prediction of nitrous oxide (N_2O) production in a nitrifying sequencing batch reactor. *Process Safety and Environmental Protection (Accepted for publication).*

Miao, Y., Zhang, L., Yang, Y., Peng, Y., Li, B., Wang, S. & Zhang, Q. 2016 Start-up of single-stage partial nitrification-anammox process treating low-strength sewage and its restoration from nitrate accumulation. *Bioresource Technology* **218**, 771–779. doi:10.1016/j.biortech.2016.06.125.

Mokhayeri, Y., Riffat, R., Murthy, S., Bailey, W., Takacs, I. & Bott, C. 2009 Balancing yield, kinetics and cost for three external carbon sources used for suspended growth post-denitrification. *Water Science and Technology* **60** (10), 2485–2491. doi:10.2166/wst.2009.623.

Morales, N., Río, Á. V., Vázquez-Padín, J. R., Méndez, R., Mosquera-Corral, A. & Campos, J. L. 2015 Integration of the anammox process to the rejection water and main stream lines of WWTPs. *Chemosphere* **140**, 99–105. doi:10.1016/j.chemosphere.2015.03.058.

Mozumder, M. S., Picioreanu, C., van Loosdrecht, M. C. M. & Volcke, E. I. P. 2014 Effect of heterotrophic growth on autotrophic nitrogen removal in a granular sludge reactor. *Environmental Technology* **35** (5–8), 1027–1037. doi:10.1080/09593330.2013.859711.

Mulder, J. W., van Loosdrecht, M. C. M., Hellinga, C. & van Kempen, R. 2001 Full-scale application of the SHARON process for treatment of rejection water of digested sludge dewatering. *Water Science and Technology* **43** (11), 127–134. https://doi.org/10.2166/wst.2001.0675.

Muñoz, A. C. 2020 *Mainstream Deammonification Process Monitoring by Bacterial Activity Tests. Dissertation*, KTH Royal Institute of Technology, Stockholm, Sweden. Available from: http://urn.kb.se/resolve?urn=urn:nbn:se:kth:diva-281698

Ni, B. J. & Yu, H. Q. 2008 An approach for modeling two-step denitrification in activated sludge systems. *Chemical Engineering Science* **63** (3), 1449–1459. doi:10.1016/j.ces.2007.12.003.

Ni, B.-J., Ruscalleda, M. & Smeths, B. 2012 Evaluation on the microbial interactions of anaerobic ammonium oxidizers and heterotrophs in anammox biofilm. *Water Research* **46** (15), 4645–4652. doi:10.1016/j.watres.2012.06.016.

Ni, B.-J., Peng, L., Law, Y., Guo, J. & Yuan, Z. 2014 Modeling of nitrous oxide production by autotrophic ammonia-oxidizing bacteria with multiple production pathways. *Environmental Science and Technology* **48** (7), 3916–3924. doi:10.1021/es405592 h.

Nogueira, R. & Melo, L. F. 2006 Competition between *Nitrospira* spp. and *Nitrobacter* spp. in nitrite-oxidizing bioreactors. *Biotechnology and Bioengineering* **95** (1), 169–175. doi:10.1002/bit.21004.

O'Shaughnessy, M. 2016 *Mainstream Deammonification*. Water Environment Federation, IWA Publishing. https://iwaponline.com/ebooks/book/301/Mainstream-Deammonification?redirectedFrom=PDF.

Ostace, S. G., Cristea, V. M. & Agachi, P. Ş. 2011 Extension of activated sludge model no 1 with two-step nitrification and denitrification processes for operation improvement. *Environmental Engineering and Management Journal* **10** (10), 1529–1544. doi:10.30638/eemj.2011.214.

Pan, Y., Ni, B.-J. & Yuan, Z. 2013 Modeling electron competition among nitrogen oxides reduction and N2O accumulation in denitrification. *Environmental Science & Technology* **47** (19), 11083–11091. doi:10.1021/es402348n.

Pan, Y., Ni, B.-J., Lu, H., Chandran, K., Richardson, D. J. & Yuan, Z. 2015 Evaluating two concepts for the modelling of intermediates accumulation during biological denitrification in wastewater treatment. *Water Research* **71**, 21–31. doi:10.1016/j.watres.2014.12.029.

Pan, Y., Ni, B.-J., Liu, Y. & Guo, J. 2016 Modeling of the interaction among aerobic ammonium-oxidizing archaea/bacteria and anaerobic ammonium-oxidizing bacteria. *Chemical Engineering Science* **150**, 35–40. doi:10.1016/j.ces.2016.05.002.

Peng, L., Xie, Y., van Beeck, W., Zhu, W., van Tendeloo, M., Tytgat, T., Lebeer, S. & Vlaeminck, S. E. 2020 Return-sludge treatment with endogenous free nitrous acid limits nitrate production and N2O emission for mainstream partial nitritation/Anammox. *Environmental Science and Technology* **54** (9), 5822–5831. doi:10.1021/acs.est.9b06404.

Pérez, J., Lotti, T., Kleerebezem, R., Picioreanu, C. & van Loosdrecht, M. C. M. 2014 Outcompeting nitrite-oxidizing bacteria in single-stage nitrogen removal in sewage treatment plants: a model-based study. *Water Research* **66**, 208–218. doi:10.1016/j.watres.2014.08.028.

Pérez, J., Isanta, E. & Carrera, J. 2015 Would a two-stage N-removal be a suitable technology to implement at full scale the use of anammox for sewage treatment? *Water Science and Technology* **72** (6), 858–865. doi:10.2166/wst.2015.281.

Pérez, J., Laureni, M., van Loosdrecht, M. C. M., Persson, F. & Gustavsson, D. J. I. 2020 The role of the external mass transfer resistance in nitrite oxidizing bacteria repression in biofilm-based partial nitritation/anammox reactors. *Water Research* **186**, 116348. doi:10.1016/j.watres.2020.116348.

Pocquet, M., Wu, Z., Queinnec, I. & Spérandio, M. 2016 A two pathway model for N_2O emissions by ammonium oxidizing bacteria supported by the NO/N_2O variation. *Water Resource* **88**, 948–959. doi:10.1016/j.watres.2015.11.029.

Poot, V., Hoekstra, M., Geleijnse, M. A., van Loosdrecht, M. C. M. & Pérez, J. 2016 Effects of the residual ammonium concentration on NOB repression during partial nitritation with granular sludge. *Water Research* **106**, 518–530. doi:10.1016/j.watres.2016.10.028.

Qian, W., Ma, B., Li, X., Zhang, Q. & Peng, Y. 2019 Long-term effect of pH on denitrification: high pH benefits achieving partial-denitrification. *Bioresource Technology* **278**, 444–449. doi:10.1016/j.biortech.2019.01.105.

Regmi, P., Miller, M. W., Holgate, B., Bunce, R., Park, H., Chandran, K., Wett, B., Murthy, S. & Bott, C. B. 2014 Control of aeration, aerobic SRT and COD input for mainstream nitritation/denitritation. *Water Research* **57** (15), 162–171. doi:10.1016/j.watres.2014.03.035.

Regmi, P., Holgate, B., Fredericks, D., Miller, M. W., Wett, B., Murthy, S. & Bott, C. 2015a Optimization of a mainstream nitritation-denitritation process and anammox polishing. *Water Science and Technology* **72** (4), 632–642. doi:10.2166/wst.2015.261.

Regmi, P., Holgate, B., Miller, M. W., Park, H., Chandran, K., Wett, B., Murthy, S. & Bott, C. B. 2015b Nitrogen polishing in a fully anoxic anammox MBBR treating mainstream nitritation-denitritation effluent. *Biotechnology and Bioengineering* **113** (3), 635–642. doi:10.1002/bit.25826.

Revollar, S., Vilanova, R., Vega, P., Francisco, M. & Meneses, M. 2020 Wastewater treatment plant operation: simple control schemes with a holistic perspective. *Sustainability* **12** (3), 768–796. doi:10.3390/su12030768.

Ribera-Guardia, A., Kassotaki, E., Gutierrez, O. & Pijuan, M. 2014 Effect of carbon source and competition for electrons on nitrous oxide reduction in a mixed denitrifying microbial community. *Process Biochemistry* **49** (12), 2228–2234. doi:10.1016/j.procbio.2014.09.020.

Rieger, L., Gillot, S., Langergraber, G., Ohtsuki, T., Shaw, A., Takacs, I. & Winkler, S. 2012 *Guidelines for Using Activated Sludge Models*. IWA Publishing, London, UK. eISBN:9781780401164.

Riet, K. V. & Lans, R. V. 2011 2.07 - Mixing in Bioreactor Vessels. In: *Comprehensive Biotechnology*, 2nd edn. Academic Press, pp. 63–80. doi:10.1016/B978-0-08-088504-9.00083-0.

Roots, P., Sabba, F., Rosenthal, A. F., Wang, Y., Yuan, Q., Rieger, L., Yang, F., Kozak, J. A., Zhang, H. & Wells, G. F. 2020 Integrated shortcut nitrogen and biological phosphorus removal from mainstream wastewater: process operation and modeling. *Environmental Science: Water Research & Technology* **6** (3), 566–580. doi:10.1039/C9EW00550A.

Rosenthal, A., Schraa, O., Rieger, L., Zhang, H., Kozak, J., Yang, F., Roots, F. & Wells, G. F. 2018 Simulation of dissolved oxygen-and ammonia-based aeration control strategies in a mainstream deammonification biofilm process. In: *Proceedings of the Water Environment Federation*. Water Environment Federation, pp. 5238–5247. doi:10.2175/193864718825138501.

Rubio-Rincón, F. J., Lopez-Vazquez, C. M., Welles, L., van Loosdrecht, M. C. M. & Brdjanovic, D. 2017 Cooperation between *Candidatus* Competibacter and *Candidatus* Accumulibacter clade I, in denitrification and phosphate removal processes. *Water Research* **120**, 156–164. doi:10.1016/j.watres.2017.05.001.

Sadowski, M., Regmi, P., Wett, B., Murthy, S. & Bott, C. 2015 Comparison of aeration strategies for optimization of nitrogen removal in an A/B process: DO, ABAC, and AvN control. In *Proceedings of the Water Environment Federation, WEFTEC 2015*. Water Environment Federation, pp. 4905–4916. doi:10.2175/193864715819538633.

Salem, S., Berends, D., Heijnen, J. & van Loosdrecht, M. C. M. 2002 Model-based evaluation of a new upgrading concept for N-removal. *Water Science and Technology* **45** (6), 169–176. PMID: 11989870.

Schielke-Jenni, S., Villez, K., Morgenroth, E. & Udert, K. 2015 Observability of anammox activity in single-stage nitritation/anammox reactors using mass balances. *Environmental Science: Water Research & Technology* **1** (4), 523–534. https://doi.org/10.1039/C5EW00045A.

Schoepflin, S., Macmanus, J., Chengua, L., McCullough, K., Klaus, S., De Clippeleir, H. & Wilson, C. Manuscript in preparation *Startup Strategies for Mainstream Anammox Polishing in Moving Bed Biofilm Reactors (MBBRs)*.

Seuntjens, D., Han, M., Kerckhof, F.-M., Boon, N., Al-Omari, A., Takacs, I., Meerburg, F., De Mulder, C., Wett, B., Bott, C., Murthy, S., Arroyo, J. M. C., De Clippeleir, H. & Vlaeminck, S. E. 2018 Pinpointing wastewater and process parameters controlling the AOB to NOB activity ratio in sewage treatment plants. *Water Research* **138**, 37–46. doi:10.1016/j.watres.2017.11.044.

Shao, Q., Wan, F., Du, W. & He, J. 2021 Enhancing biological nitrogen removal for a retrofit project using wastewater with a low C/N ratio – a model-based study. *Environmental Science and Pollution Research* **28**, 53074–53086. doi:10.1007/s11356-021-14396-2.

Shaw, A., Takács, I., Pagilla, K. R. & Murthy, S. 2013 A new approach to assess the dependency of extant half-saturation coefficients on maximum process rates and estimate intrinsic coefficients. *Water Research* **47** (16), 5986–5994. doi:10.1016/j.watres.2013.07.003.

Shi, L., Du, R. & Peng, Y. 2019 Achieving partial denitrification using carbon sources in domestic wastewater with waste-activated sludge as inoculum. *Bioresource Technology* **283**, 18–27. doi:10.1016/j.biortech.2019.03.063.

Shourjeh, M., Kowal, P., Lu, X., Xie, L. & Drewnowski, J. 2021 Development of strategies for AOB and NOB competition supported by mathematical modeling in terms of successful deammonification implementation for energy-efficient WWTPs. *Processes* **9** (3), 562–587. doi:10.3390/pr9030562.

Siegrist, H., Salzgeber, D., Eugster, J. & Joss, A. 2008 Anammox brings WWTP closer to energy autarky due to increased biogas production and reduced aeration energy for N-removal. *Water Science and Technology* **57** (3), 383–388. doi:10.2166/wst.2008.048.

Sin, G., Kaelin, D., Kampschreur, M. J., Takács, I., Wett, B., Gernaey, K. V., Reiger, L., Siegrist, H. & van Loosdrecht, M. C. M. 2008 Modelling nitrite in wastewater treatment systems: a discussion of different modelling concepts. *Water Science and Technology* **58** (6), 1155–1171. doi:10.2166/wst.2008.485.

Soler-Jofra, A., Wang, R., Kleerebezem, R., van Loosdrecht, M. C. M. & Pérez, J. 2019 Stratification of nitrifier guilds in granular sludge in relation to nitritation. *Water Research* **148**, 479–491. doi:10.1016/j.watres.2018.10.064.

Spérandio, M., Pocque, M., Guo, L., Ni, B.-J., Vanrolleghem, P. A. & Yuan, Z. 2016 Evaluation of different nitrous oxide production models with four continuous long-term wastewater treatment process data series. *Bioprocess and Biosystem Engineering* **39** (3), 493–510. doi:10.1007/s00449-015-1532-2.

Staunton, E. T. & Aitken, M. D. 2015 Coupling nitrogen removal and anaerobic digestion for energy recovery from swine waste: 2 nitration/Anammox. *Environmental Engineering Science* **32** (9), 750–760. doi:10.1089/ees.2015.0063.

Strous, M., Heijnen, J., Kuenen, J. & Jetten, M. 1998 The sequencing batch reactor as a powerful tool for the study of slowly growing anaerobic ammonium-oxidizing microorganisms. *Applied Microbiology and Biotechnology* **50**, 589–596. doi:10.1007/s002530051340.

Tao, C. & Hamouda, M. A. 2019 Steady-state modeling and evaluation of partial nitrification-anammox (PNA) for moving bed biofilm reactor and integrated fixed-film activated sludge processes treating municipal wastewater. *Journal of Water Process Engineering* **31**, 100854. doi:10.1016/j.jwpe.2019.100854.

Trojanowicz, K., Plaza, E. & Trela, J. 2019 Model extension, calibration and validation of partial nitritation-anammox process in moving bed biofilm reactor (MBBR) for reject and mainstream wastewater. *Environmental Technology* **40** (4), 1079–1100. doi:10.1080/09593330.2017.1397765.

Valverde Pérez, B., Mauricio Iglesias, M. & Sin, G. 2016 Systematic design of an optimal control system for the SHARON-Anammox process. *Journal of Process Control.* **39** (1), pp. 1–10. doi: 10.1016/j.jprocont.2015.12.009.

Van Dongen, U., Jetten, M. S. M. & van Loosdrecht, M. C. M. 2001 The SHARON®-Anammox® process for treatment of ammonium rich wastewater. *Water Science and Technology* **44** (1), 153–160.

Van Hulle, S. W. H., Volcke, E. I. P., López Teruel, J., Donckels, B., van Loosdrecht, M. C. M. & Vanrolleghem, P. A. 2007 Influence of temperature and pH on the kinetics of the Sharon nitritation process. *Journal of Chemical Technology and Biotechnology* **82** (5), 471–480. doi:10.1002/jctb.1692.

Vangsgaard, A. K., Mutlu, A. G., Gernaey, K. V., Smets, B. F. & Sin, G. 2013 Calibration and validation of a model describing complete autotrophic nitrogen removal in a granular SBR system. *Journal of Chemical Technology and Biotechnology* **88**, 2007–2015. doi:10.1002/jctb.4060.

Vanrolleghem, P. A., Mannina, G., Cosenza, A. & Neumann, M. B. 2015 Global sensitivity analysis for urban water quality modelling: terminology, convergence and comparison of different methods. *Journal of Hydrology* **522**, 339–352. doi:10.1016/j.jhydrol.2014.12.056.

Vasilaki, V., Massara, T. M., Stanchev, P., Fatone, F. & Katsou, E. 2019 A decade of nitrous oxide (N2O) monitoring in full-scale wastewater treatment processes: a critical review. *Water Research* **161**, 392–412. doi:10.1016/j.watres.2019.04.022.

Volcke, E. I. P., Van Hulle, S. W. H., Donckels, B. M. R., van Loosdrecht, M. C. M. & Vanrolleghem, P. A. 2005 Coupling the SHARON process with anammox: model-based scenario analysis with focus on operating costs. *Water Science and Technology* **52** (4), 107–115. doi:10.2166/wst.2005.0093

Volcke, E. I. P., Gernaey, K. V., Vrecko, D., Jeppsson, U., van Loosdrecht, M. C. M. & Vanrolleghem, P. A. 2006 Plant-wide (BSM2) evaluation of reject water treatment with a SHARON-Anammox process. *Water Science and Technology* **54** (8), 93–100. doi:10.2166/wst.2006.822.

Volcke, E. I. P., Sanchez, O., Steyer, J. P., Dabert, P. & Bernert, N. 2008 Microbial population dynamics in nitrifying reactors: experimental evidence explained by a simple model including interspecies competition. *Process Biochemistry* **43** (2), 1398–1406. doi:10.1016/j.procbio.2008.08.013.

Volcke, E. I. P., Picioreanu, C., De Baets, B. & van Loosdrecht, M. C. M. 2010 Effect of granule size on autotrophic nitrogen removal in a granular sludge reactor. *Environmental Technology* **31** (11), 1271–1280. doi:10.1080/09593331003702746.

Volcke, E. I. P., Picioreanu, C., De Baets, B. & van Loosdrecht, M. C. M. 2012 The granule size distribution in an anammox-based granular sludge reactor affects the conversion – implications for modeling. *Biotechnology and Bioengineering* **109** (7), 1629–1636. doi:10.1002/bit.24443.

Von Schulthess, R., Kühni, M. & Gujer, W. 1995 Release of nitric and nitrous oxides from denitrifying activated sludge. *Water Research* **29** (1), 215–226. doi:10.1016/0043-1354(94)E0108-I.

Von Stosch, M., Oliveira, R., Peres, J. & de Azevedo, S. F. 2014 Hybrid semi-parametric modeling in process systems engineering: past, present and future. *Computers and Chemical Engineering* **60**, 86–101. doi:10.1016/j.compchemeng.2013.08.008.

Wade, M. J. & Wolkowicz, G. S. K. 2021 Bifurcation analysis of an impulsive system describing partial nitritation and anammox in a hybrid reactor. *Environmental Science & Technology* **55** (3), 2099–2109. doi:10.1021/acs.est.0c06275.

Wan, X., Baeten, J. E. & Volcke, E. I. P. 2019 Effect of operating conditions on N₂O emissions from one-stage partial nitritation-anammox reactors. *Biochemical Engineering Journal* **143**, 24–33. doi:10.1016/j.bej.2018.12.004.

Wang, D., Wang, Q., Laloo, A., Xu, Y., Bond, P. L. & Yuan, Z. 2016 Achieving stable nitritation for mainstream deammonification by combining free nitrous acid-based sludge treatment and oxygen limitation. *Scientific Reports* **6**, 25547. doi:10.1038/srep25547.

Wang, H., Xu, G., Qiu, Z., Zhou, Y. & Liu, Y. 2019 NOB suppression in pilot-scale mainstream nitritation-denitritation system coupled with MBR for municipal wastewater treatment. *Chemosphere* **216**, 633–639. doi:10.1016/j.chemosphere.2018.10.187.

Wang, Q., Ni, B.-J., Lemaire, R., Hao, X. & Yuan, Z. 2016 Modeling of nitrous oxide production from nitritation reactors treating real anaerobic digestion liquor. *Scientific Reports* **6**, 25336. doi:10.1038/srep25336.

Wang, X., Zhao, J., Yu, D., Du, S., Yuan, M. & Zhen, J. 2019 Evaluating the potential for sustaining mainstream anammox by endogenous partial denitrification and phosphorus removal for energy-efficient wastewater treatment. *Bioresource Technology* **284**, 302–314. doi:10.1016/j.biortech.2019.03.127.

Wett, B. 2007 Development and implementation of a robust deammonification process. *Water Science and Technology* **56** (7), 81–88. doi:10.2166/wst.2007.611.

Wett, B. & Rauch, W. 2003 The role of inorganic carbon limitation in biological nitrogen removal of extremely ammonia concentrated wastewater. *Water Research* **37** (5), 1100–1110. doi:10.1016/s0043-1354(02)00440-2.

Wett, B., Nyhuis, G., Takács, I. & Murthy, S. 2010 Development of enhanced deammonification selector. In *Proceedings of the Water Environment Federation, WEFTEC 2010*. Water Environment Federation, pp. 5917–5926. doi:10.2175/193864710798194139.

Wett, B., Podmirseg, S. M., Gómez-Brandón, M., Hell, M., Nyhuis, G., Bott, C. & Murthy, S. 2015 Expanding DEMON sidestream deammonification technology towards mainstream application. *Water Environment Research* **87** (12), 2084–2089. doi:10.2175/106143015X14362865227319.

Wild, D., Von Schulthess, R. & Gujer, W. 1995 Structured modelling of denitrification intermediates. *Water Science and Technology* **31** (2), 45–54. doi:10.2166/wst.1995.0070.

Wu, J. 2017 Comparison of control strategies for single-stage partial nitrification-anammox granular sludge reactor for mainstream sewage treatment – a model-based evaluation. *Environmental Science and Pollution Research* **24**, 25839–25848. doi:10.1007/s11356-017-0230-9.

Wyffels, S., Van Hulle, S. W. H., Boeckx, P., Volcke, E. I. P., Van Cleemput, O., Vanrolleghem, P. A. & Verstraete, W. 2004 Modeling and simulation of oxygen-limited partial nitritation in a membrane-assisted bioreactor (MBR). *Biotechnology and Bioengineering* **86** (5), 531–542. doi:10.1002/bit.20008.

Xu, G., Wang, H., Gu, J., Shen, N., Qiu, Z., Zhou, Y. & Liu, Y. 2017 A novel A-B process for enhanced biological nutrient removal in municipal wastewater reclamation. *Chemosphere* **189**, 39–45. doi:10.1016/j.chemosphere.2017.09.049.

Yang, Q., Shen, N., Lee, Z. M.-P., Xu, G., Cao, Y., Kwok, B., Lay, W., Liu, Y. & Zhou, Y. 2016 Simultaneous nitrification, denitrification and phosphorus removal (SNDPR) in a full-scale water reclamation plant located in warm climate. *Water Science and Technology* **74** (2), 448–456. doi:10.2166/wst.2016.214.

You, J., Das, A., Dolan, E. & Hu, Z. 2009 Ammonia-oxidizing archaea involved in nitrogen removal. *Water Research* **43** (7), 1801–1809. doi:10.1016/j.watres.2009.01.016.

You, Q. G., Wang, J. H., Qi, G. X., Zhou, Y. M., Guo, Z. W., Shen, Y. & Gao, X. 2020 Anammox and partial denitrification coupling: a review. *RSC Advances* **10** (21), 12554–12572. doi:10.1039/D0RA00001A.

Yu, L.-F., Du, Q.-Q., Fu, X.-T., Zhang, R., Li, W.-J. & Peng, D.-C. 2016 Community structure and activity analysis of the nitrifiers in raw sewage of wastewater treatment plants. *Huan Jing Ke Xue* **37** (11), 4366–4371. doi:10.13227/j.hjkx.201605026.

Zekker, I., Rikmann, E., Tenno, T., Saluste, A., Tomingas, M., Menert, A., Loorits, L., Lemmiksoo, V. & Tenno, T. 2012 Achieving nitritation and anammox enrichment in a single moving-bed biofilm reactor treating reject water. *Environmental Technology* **33** (4–6), 703–710. doi:10.1080/09593330.2011.588962.

Zhang, M., Wang, S., Ji, B. & Liu, Y. 2019 Towards mainstream deammonification of municipal wastewater: partial nitrification-anammox versus partial denitrification-anammox. *Science of the Total Environment* **692**, 394–400. doi:10.1016/j.scitotenv.2019.07.293.

Zhang, M., Li, N., Chen, W. & Wu, J. 2020 Steady-state and dynamic analysis of the single-stage anammox granular sludge reactor show that bulk ammonium concentration is a critical control variable to mitigate feeding disturbances. *Chemosphere* **251**, 126361. doi:10.1016/j.chemosphere.2020.126361.

Zhang, W., Peng, Y., Zhang, L., Li, X. & Zhang, Q. 2020 Simultaneous partial nitritation and denitritation coupled with polished anammox for advanced nitrogen removal from low C/N domestic wastewater at low dissolved oxygen conditions. *Bioresource Technology* **305**, 123045. doi:10.1016/j.biortech.2020.123045.

Zhang, Z., Zhang, Y. & Chen, Y. 2020 Recent advances in partial denitrification in biological nitrogen removal: from enrichment to application. *Bioresource Technology* **298**. doi:10.1016/j.biortech.2019.122444.

Zhang, X., Yu, B., Zhang, N., Zhang, H., Wang, C. & Zhang, H. 2016 Effect of inorganic carbon on nitrogen removal and microbial communities of CANON process in a membrane bioreactor. *Bioresource Technology* **202**, 113–118. doi:10.1016/j.biortech.2015.11.083.

Zhao, Y., Miao, J., Ren, X. & Wu, G. 2018 Effect of organic carbon on the production of biofuel nitrous oxide during the denitrification process. *International Journal of Environmental Science and Technology* **15**, 461–470. doi:10.1007/s13762-017-1397-9.

First received 14 January 2022; accepted in revised form 7 April 2022. Available online 20 April 2022

doi: 10.2166/wst.2022.048

An influent generator for WRRF design and operation based on a recurrent neural network with multi-objective optimization using a genetic algorithm

Feiyi Li [a,b,*] and Peter A. Vanrolleghem [a,b]

[a] modelEAU, Université Laval, 1065, Avenue de la Médecine, Québec, QC G1 V 0A6, Canada
[b] CentrEau, Québec Water Research Center, 1065 avenue de la Médecine, Québec, QC G1 V 0A6, Canada
*Corresponding author. E-mail: feiyi.li.1@ulaval.ca

FL, 0000-0003-4278-730X; PAV, 0000-0003-1695-1313

ABSTRACT

Nowadays, modelling, automation and control are widely used for Water Resource Recovery Facilities (WRRF) upgrading and optimization. Influent generator (IG) models are used to provide relevant input time series for dynamic WRRF simulations used in these applications. Current IG models found in literature are calibrated on the basis of a single performance criterion, such as the mean percentage error or the root mean square error. This results in the IG being adequate on average but with a lack of representativeness of, for instance, the observed temporal variability of the dataset. However, adequately capturing influent variability may be important for certain types of WRRF optimization, e.g., reaction to peak loads, control system performance evaluation, etc. Therefore, in this study, a data-driven IG model is developed based on the long short-term memory (LSTM) recurrent neural network and is optimized by a multi-objective genetic algorithm for both mean percentage error and variability. Hence, the influent generator model is able to generate a time series with a probability distribution that better represents reality, thus giving a better influent description for WRRF design and operation. To further increase the variability of the generated time series and in this way approximate the true variability better, the model is extended with a random walk process.

Key words: digitalization, LSTM, machine learning, NSGA-II, variability, wastewater modelling

HIGHLIGHTS

- A data-driven influent generator based on the long short-term memory (LSTM) is proposed.
- To represent both the average behaviour and variability of the influent, the IG is trained by a multi-objective genetic algorithm.
- The model is extended with a random walk stochastic process with a probability distribution that better represents reality, giving a better influent description for model-based WRRF design and operation.

1. INTRODUCTION AND BACKGROUND

Nowadays, modelling, automation and control are widely used for Water Resource Recovery Facilities (WRRF) upgrading and optimization. However, because of a lack of adequate input datasets, their full potential remains underexploited. For example, for WRRF design, engineers usually make initial sizing by using design guidelines based on average loads and safety factors (Talebizadeh *et al.* 2015). Also, controller performance can benefit from influent forecasting; for example, with nitrogen load data for ammonia-based aeration control (Newhart *et al.* 2020). Thus, a reliable influent generator model is becoming increasingly necessary as input to WRRF modelling and process control studies (Gernaey *et al.* 2011; Martin & Vanrolleghem 2014).

Many influent generators (IG) have already been presented in literature. They can be organized into two categories: phenomenological models (Flores-Alsina *et al.* 2014; Langeveld *et al.* 2017) and data-driven models (Ahnert *et al.* 2016; Borzooei *et al.* 2019; Li *et al.* 2020).

Compared with phenomenological models, data-driven models usually require less calculation effort thanks to the lower model complexity, while still delivering a high-performance result. As a powerful data-driven tool, machine learning approaches have been increasingly used in water engineering thanks to the increasing data collection and data mining tool development (Corominas *et al.* 2018). Recently, Artificial Neural Networks (ANNs) have been studied for WWTP influent generation, especially for short-term flow forecasting and for pollutant concentration prediction (El-Din & Smith 2002; Aminabad *et al.* 2013; Ma *et al.* 2014; Shokry *et al.* 2018; Li *et al.* 2020).

The first challenge of data-driven IG development is that despite the fact that ANNs have been proven to have an excellent ability for modelling water systems, it is necessary to first define an adequate ANN model architecture. Since the wastewater generation process of a sewer catchment is a complex nonlinear process and it depends on historical information (e.g., previous rain events), the recurrent neural network (RNN) is expected to be the most appropriate architecture as it allows learning time-sequential behaviour thanks to its 'internal memory'. RNNs are found to be better than fully connected neural networks, such as multilayer perceptrons (MLP). The long short-term memory (LSTM) is one type of RNN that was proven to have excellent performance for time series modelling, especially when learning long-term dependencies (Hochreiter & Schmidhuber 1997). It can therefore be expected to provide good performance for IG modelling.

The second challenge of influent generator development is that even though multiple influent generators have been presented in literature, the calibration process used is single objective optimization; in other words, IGs are calibrated with only one performance criterion in mind, typically defined as either the mean percentage error (MPE) or the root mean square error (RMSE). These criteria lead to IG that are adequate on average but they may not represent the observed variability well. In other words, without better consideration of the statistical distribution of the data, the model will not be able to fully capture the influent variability. Representing variability well is important to better define WRRF design parameters, such as the expected maximum load and hydraulic capacity and for scenarios analysis in which one studies the influent disturbance impact when evaluating operation under peak conditions, etc.

To solve these issues with existing IG models, a novel data-driven IG model is developed using the LSTM to describe wastewater quality. The model is calibrated by a multi-objective genetic algorithm (MoGA) (Fonseca & Fleming 1999), which next to being able to handle multiple criteria also outperforms traditional gradient descent optimization tools in dealing with local minima, ensuring efficient optimization during IG development (Rojas 1996).

The provided case study illustrates that such an IG model is able to generate long-term concentration data for COD and TSS, based only on previous flowrate and weather data. It has good performance in terms of mean percentage error, while at the same time providing excellent similarity to the variability of the real dataset.

2. CASE STUDY DESCRIPTION AND DATA PRE-TREATMENT

In this study, the IG modelling approach was developed based on data collected in Quebec City, Canada (Tik & Vanrolleghem 2017). The catchment is a combined sewer system, with a very variable flowrate and pollutant concentration under different weather conditions, such as storm water and particularly, the snowmelt at the end of winter. The Quebec City WRRF is a carbon removing treatment plant and its dataset includes weather information (rain and temperature), daily flowrate, and daily COD and TSS concentrations.

The influent flow rate and concentrations show recurring hourly and daily variations but are disturbed by rain events. Data pre-treatment was conducted according the procedure of Alferes *et al.* (2013), including outlier removal, data smoothing, and a univariate fault detection method. Then the normalization process was applied to scale the data in the range between 0 and 1.

In order to remove the effect of precipitation, signal noise, and obtain a typical long-term yearly dry weather pattern, a Chebyshev bandpass filter (2nd order, Type I) was selected to extract the signal at these specific frequencies from the data (Schlichthärle 2011). This signal then can represent the seasonal effect and the dry weather flow corresponding to the urban activities.

3. MATERIALS AND METHODS

3.1. Fully connected ANN model

The MLP is a class of feedforward ANN where each neuron in one layer is fully connected to the next layer (Raman & Sunilkumar 1995; El-Din *et al.* 2004; Zhang *et al.* 2019). It is one of the most used ANN architectures and is used in this study as the reference, see Figure 1(a). It is trained using the classic learning process only for minimal RMSE and using the gradient descent approach; that is, the Levenberg–Marquardt backpropagation algorithm (Levenberg 1944). The sigmoid function is selected as activation function in the hidden layer, in order to represent the nonlinearity of the influent generation process.

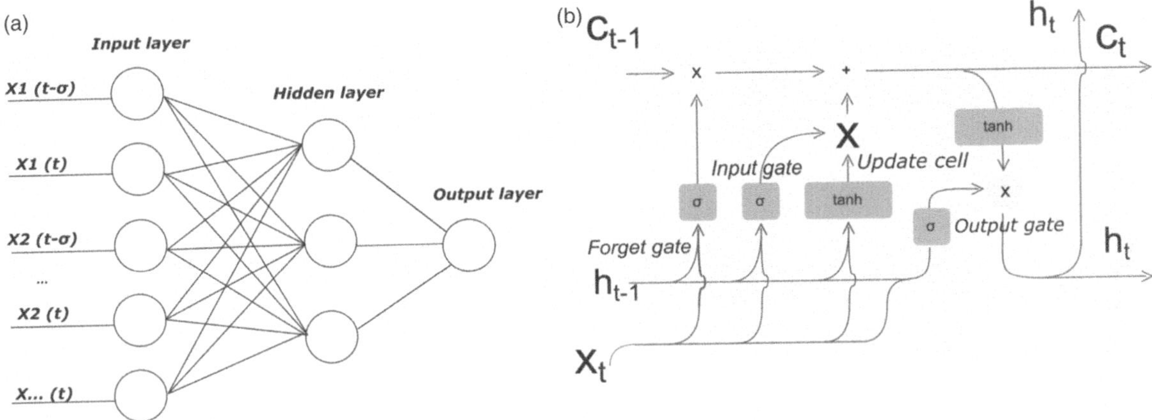

Figure 1 | Architecture for (a) MLP fully connected ANN (b) LSTM architecture.

3.2. LSTM model

Despite the fact that the MLP can learn influent dynamics, the RNN is expected to be a more adequate model structure because it can better describe the pollutant concentration that is influenced by previous conditions. To handle the long-term dependency problem, the LSTM is selected to integrate this earlier information. Figure 1(b) illustrates the architecture of LSTM with its gates structure; that is, input, forget, and output gates. The long term memory is stored and carried on by the cell state through time (Hochreiter 1998; Graves 2012). The input layers and architecture of the LSTM used in this study are presented in Figure 2.

3.3. NSGA-II method

Genetic algorithms are popular optimization approaches inspired by the process of natural selection and have quickly became a popular evolutionary algorithm (Holland 1992). It is playing an increasingly important role in machine learning and many other applications (Goldberg 1989; Ercan & Goodall 2016). Multi-objective optimization problems can be solved either by determining an entire set of optimal solutions on the Pareto front, or by combining the individual objective functions into a single one using, for instance, weights (Konak *et al.* 2006; Chiandussi *et al.* 2012). The NSGA-II method (non-dominated sorting genetic algorithm II) (Deb *et al.* 2002) is widely used in machine learning and starts to be used in the water engineering field to solve multi objective optimization problems (Yusoff *et al.* 2011; Ercan & Goodall 2016; Wang *et al.* 2019).

Figure 3 displays the flowchart of NSGA. An initial population is generated randomly. The non-dominated sorting fronts are ranked by the evaluation of their fitness. In this research, the crowding distance and tournament selection are used for the selection process (Miller & Goldberg 1996) and adaptive mutation and crossover ratios are applied to improve the search efficiency (Srinivas & Patnaik 1994; Hassanat *et al.* 2019).

Different quantitative performance criteria aim to measure how well a model simulation fits the available observations. A diversity of criteria has been studied and compared for engineering and wastewater applications (Chiandussi *et al.* 2012; Hauduc *et al.* 2015). In this study, the aim is to obtain an IG which has a high precision and, at the same time, a variability

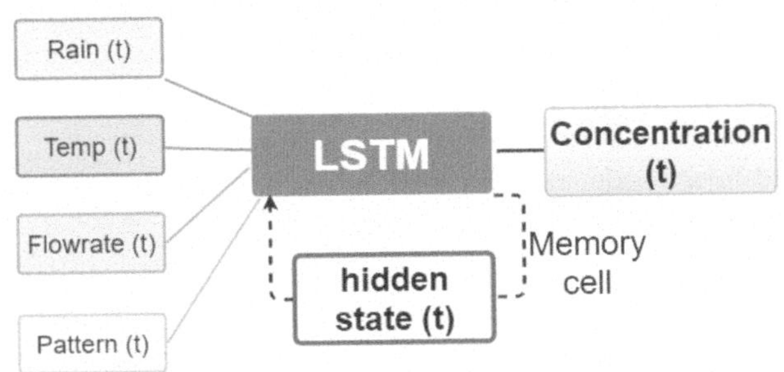

Figure 2 | The architecture of the LSTM recurrent neural network used in this study for IG modelling.

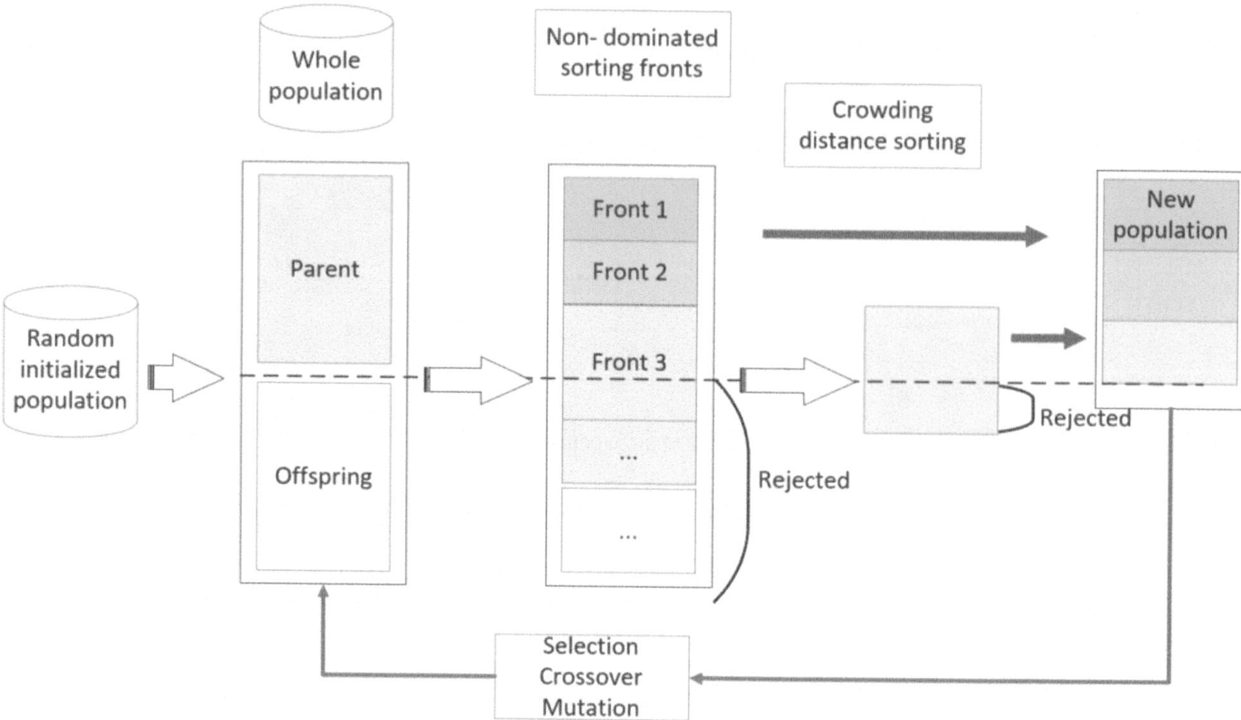

Figure 3 | NSGA-II flowchart.

similar to the one of the observed datasets. Thus, the following two criteria were chosen to represent the desired IG performance: the mean absolute percentage error (MAPE), which is a common criterion for time series, and the $KL_{divergence}$ (Kullback & Leibler 1951), which measures the difference between two probability distributions:

$$MAPE = \frac{1}{n}\sum \left| \frac{y_s - y_o}{y_s} \right| * 100\%$$

$$KL_{divergence}(P \parallel Q) = -\sum P(x) * \log\left(\frac{Q(x)}{P(x)}\right)$$

where, y_s is the simulated and y_o is the observed data, $P(x)$ is the probability distribution of the observed data and $Q(x)$ is the distribution of the simulated data.

3.4. Additional random walk

As shown in the results section, even though the IG model was optimized by NSGA-II, the result for variability was felt to be insufficient and improvement to the model structure was tried by increasing the variability it could generate. In general, a time series model is composed of a deterministic part, a stochastic part and an auto-correlated error term (Reichert & Schuwirth 2012; Villez *et al.* 2020). Therefore, by adding an error term, in this case a random walk process, to the calibrated GA model, it is expected to improve the data series variability and eliminate the autocorrelated error now present because of the insufficient inclusion of short-term dynamics. By doing this, the IG simulation is expected to achieve statistical properties closer to those of the observed data.

The stochastic process is modelled by a k-order random walk model:

$$R_t = \sum_1^k \varphi_k * R_{t-k} + \varepsilon \tag{1}$$

where R_t is a random value at time t, corresponding to the difference between the ANN model and the actual data, φ_k are the random walk model's k coefficients and ε is a white noise sequence.

4. RESULTS AND DISCUSSION

4.1. Results of LSTM-NSGA-II

The IG model developed for the case study at hand generates results for daily TSS and COD concentrations. The input includes 4 previous days of data regarding weather (rain and temperature), wastewater flow and seasonal pattern of COD, obtained after data pre-treatment. Indeed, the auto-correlation analysis shows that lags up to 4 days back had a significant autocorrelation.

The final model consists of two LSTM hidden layers and one output layer. Based on previous experience, the following settings of the NSGA-II algorithm were adopted: in order to find the global optimal solution, a population of 2,000 individuals was initialized at the beginning of the NSGA-II optimization process, and 25 search iterations were performed. The initial crossover rate and mutation rate were defined as 0.75 and 0.15, respectively.

The results of LSTM-NSGA-II are compared with a classical full-connected neural network with sigmoid function in the hidden layer, trained with back-propagation (ANN-BP). Figure 4 illustrates the Pareto front of the last iteration of the NSGA optimization for generating COD concentrations as an example. The green points represent a series of optimal non-dominated solutions in which the MAPE cannot be improved without sacrificing the KL divergence or vice versa. For visual comparison, the blue triangle represents the result of the ANN-BP. The final solution (red point) was chosen by balancing the two performance criteria and is discussed below.

As shown in Figure 5, the results for the test set illustrate the good performance of the simulated time series, see Table 1. The dilution effect during wet weather caused by stormwater inflow and subsequent rain-induced infiltration can be observed. It can also be noticed that the LSTM-NSGA-II result is more variable than the time series generated by the ANN-BP model, which confirms that the LSTM is better to describe influent variability, albeit still insufficient, as discussed below. In addition, by analysing the probability density function (PDF) and the cumulative density function (CDF), the LSTM-results are very similar to reality for both COD and TSS concentrations (see Figure 6).

This model can not only be applied to generate time series of organic pollution (COD, TSS) but also for nitrogen species (such as ammonia, see Figure 7). This can be very beneficial for database gap filling, considering that in carbon removing plants such as the one under study, the ammonia concentration is usually not measured daily. The influent time series generated in this way allows enhancing the evaluation of future nitrogen removal performance, and can also contribute to modelling nutrient recovery, and so on.

Finally, the performance of the COD, TSS and ammonia concentration time series for the test set (one solution, the red point, chosen from the Pareto front of Figure 4) is summarized in Table 1 and compared to the full-connected back-propagation ANN model optimized by gradient descent (green triangle in Figure 4).

In general, the LSTM model performs better in terms of MAPE and especially KL divergency. Thanks to cell memory, the LSTM is able to learn the impact of previous phenomena, such as the rainfall derived infiltration and inflow (RDII). By ensuring the model quality in term of MAPE, while at the same time pursuing agreement on variability, the NSGA-II allows obtaining a distribution of model results similar to the real distribution, which makes the generated influent time series have more variability than the one generated by the ANN-BP model. This improvement thus better describes the observed

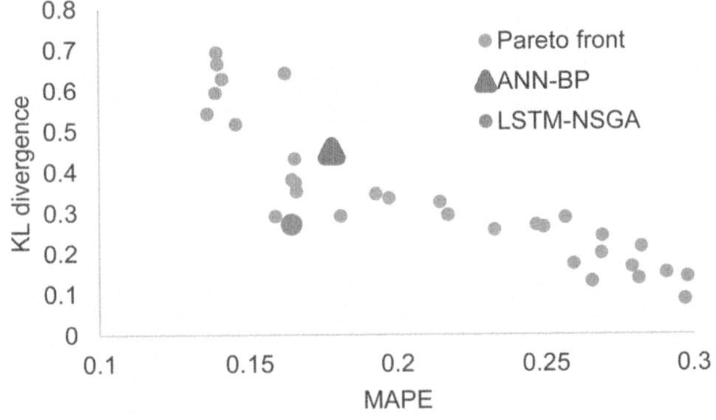

Figure 4 | NSGA-II optimal Pareto front solutions: the red point is the solution chosen for further analysis and the blue triangle represents the ANN-BP result. Please refer to the online version of this paper to see this figure in colour: http://dx.doi.org/10.2166/wst.2022.048.

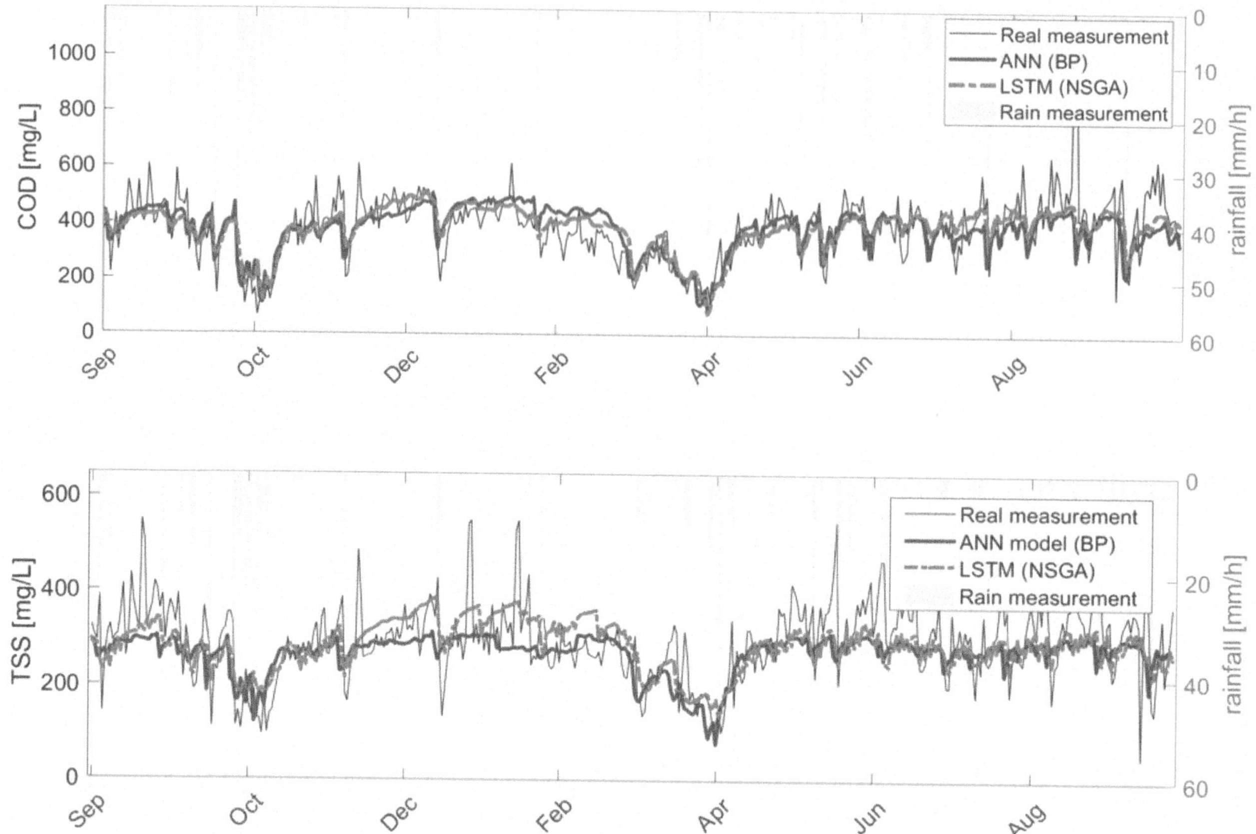

Figure 5 | COD concentrations (top) and TSS concentrations (bottom) generated by LSTM (red line) and ANN-BP (blue line), by using an input data set only including weather and flowrate data. Please refer to the online version of this paper to see this figure in colour: http://dx.doi.org/10.2166/wst.2022.048.

Table 1 | Performance analysis for LSTM (red point in Figure 4) and ANN-BP (green triangle in Figure 4) in terms of the MAPE and $KL_{divergence}$ criteria

	LSTM optimized by NSGA-II		ANN-BP	
Criteria	MAPE	$KL_{divergence}$	MAPE	$KL_{divergence}$
TSS [mg/l]	18.5%	0.28	17.7%	0.49
COD [mg/l]	16.4%	0.27	17.8%	0.45
Ammonia [mg/l]	13.7%	0.22	15.1%	0.36

randomness of the concentration time series, which is important regarding WRRF design and evaluation of WRRF control systems, for instance.

4.2. Results of benefits of adding a random walk process to the IG

As mentioned in the section Additional random walk, improvement of the IG was tried by reconstructing the difference between the model output and the observations by extending the deterministic model with a random walk process. The autocorrelation and partial autocorrelation analysis (Figure 8(a)) suggest that an order equal to 4 should be selected for the k-order random walk model. In order to represent the noise term, different distributions were compared (see Figure 8(b)), and a Weibull distribution was selected as it gave the distribution most similar to the observed error distribution.

After the autoregressive error reconstruction, Figure 9 illustrates the test set COD concentration results obtained with the model extended with a random walk process. The grey random walk band is the confidence interval for the COD

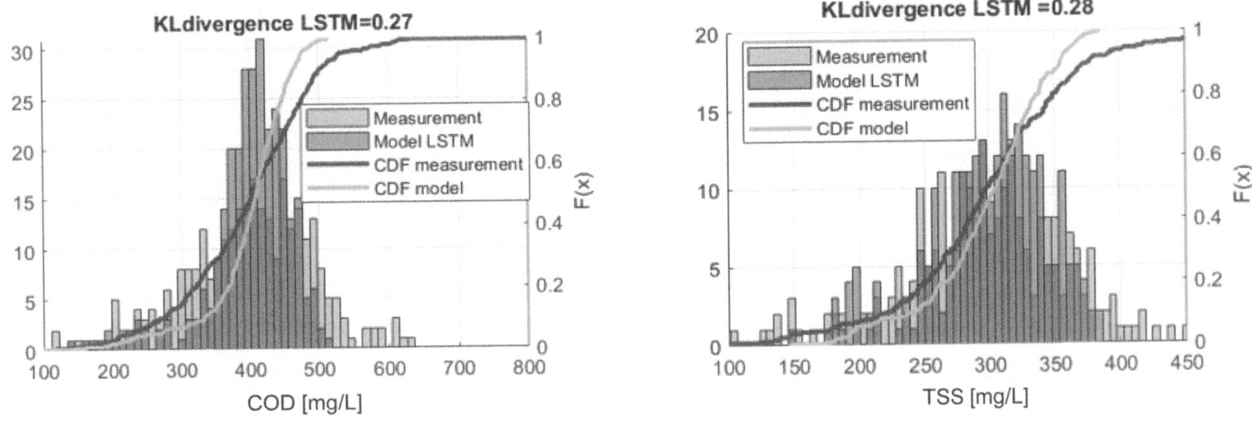

Figure 6 | PDF and CDF result for LSTM-generated COD (left) and TSS (right); the KL divergence values are given.

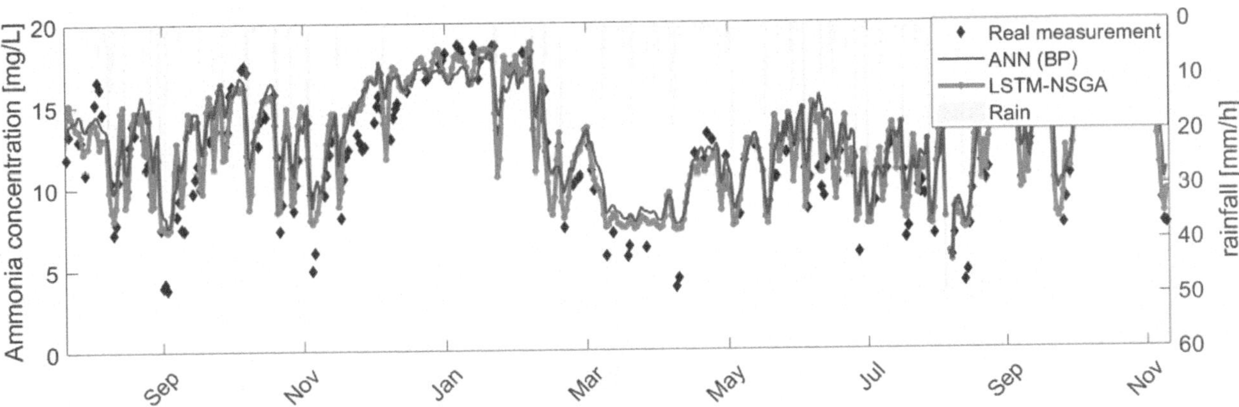

Figure 7 | Ammonia concentration time series generated by LSTM-NSGA using occasional ammonia measurements at a carbon-removing plant.

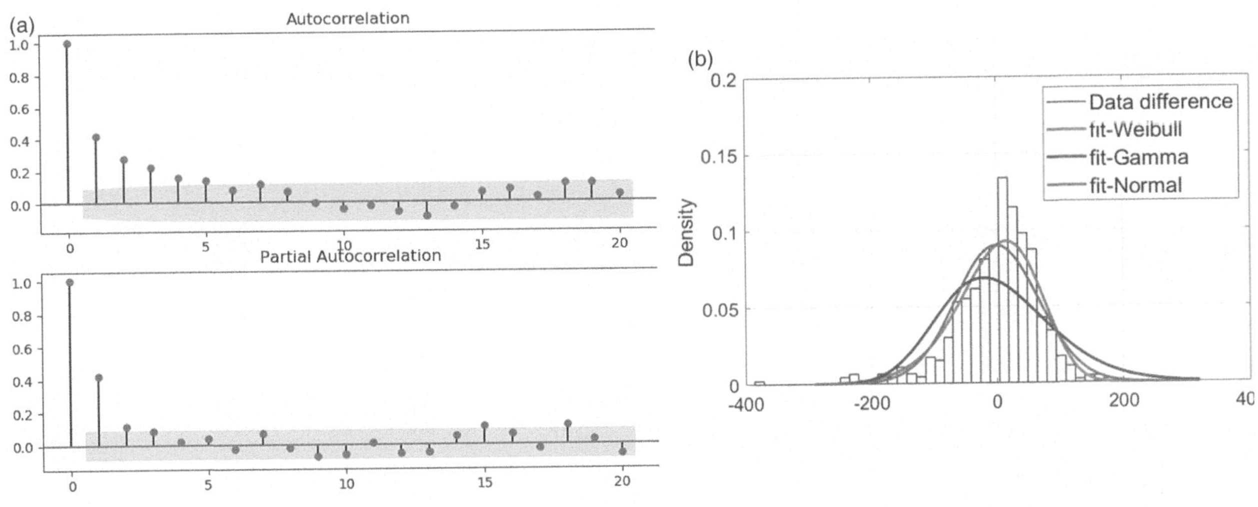

Figure 8 | (a) Autocorrelation and partial autocorrelation analysis for the error between the deterministic LSTM model output and the measured data, with the blue zone indicating the significant autocorrelation limit. (b) Fit of different distribution models to the error PDF. Please refer to the online version of this paper to see this figure in colour: http://dx.doi.org/10.2166/wst.2022.048.

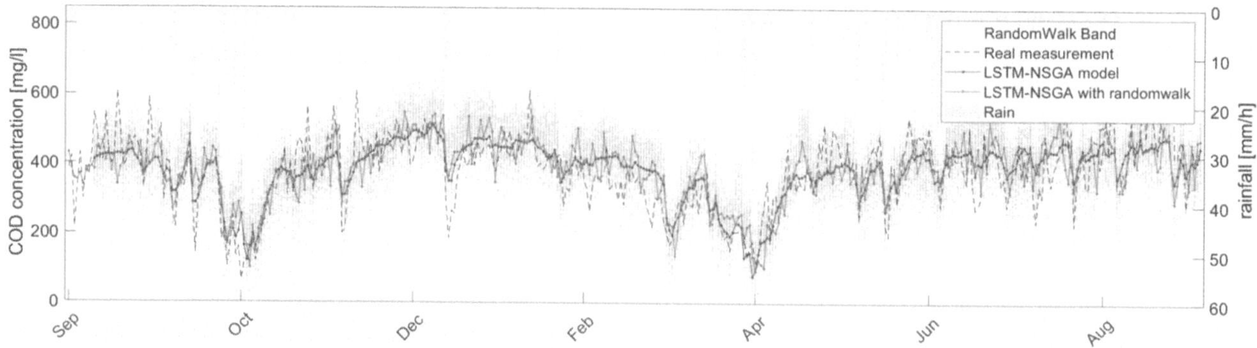

Figure 9 | COD concentration generation with the IG model extended with a random walk process. The grey band represents the confidence interval of this stochastic model generated with 200 Monte Carlo simulations.

concentrations obtained by running 200 Monte Carlo simulations with different random sequences. The red line is one example chosen among these 200 simulations. The $KL_{divergence}$ for this example decreased from 0.28 to 0.14, while MAPE increased slightly (from 16.4% to 17.5%) because of the randomness of the random walk process. It is important to highlight that the random walk process aims to re-establish the variability of the pollutants, thus representing the stochastic process in the real time series. In other words, the objective of the random walk is not to synchronize the time series with the actual measured one and, therefore, the timing of the variations is not the most essential criterion for the model selection.

The model exhibits a good result for the ammonia concentration time series as well, see Figure 10. The $KL_{divergence}$ decreased from 0.22 to 0.17 with a slight increase in MAPE (from 13.7% to 14.26%). This approach improves the ammonia concentration profile generation and database gap filling: even though the real measurement is not performed on a daily basis, the model can generate an interval for the missing data.

5. CONCLUSION

The ability to properly describe the WRRF influent variability is very crucial for a number of WRRF modelling tasks. This paper has proposed an IG model based on the LSTM Recurrent Neural Network architecture to generate daily concentration data for COD, TSS and ammonia nitrogen. Compared with a plain ANN, it results in considerably better model performance in terms of influent generation when considering both accuracy and variability simultaneously. The latter feature of the new IG-model is pursued by using a multi-objective optimization, *in casu*, the NSGA-II algorithm.

A further improvement of the variability correspondence of the generated time series was achieved by adding a random walk process to the deterministic core LSTM model. The generated time series can now achieve a distribution very similar to the one observed, and therefore, it will help deciding on WRRF design parameters, especially those related to peak hydraulic capacity and major influent disturbances, and so on.

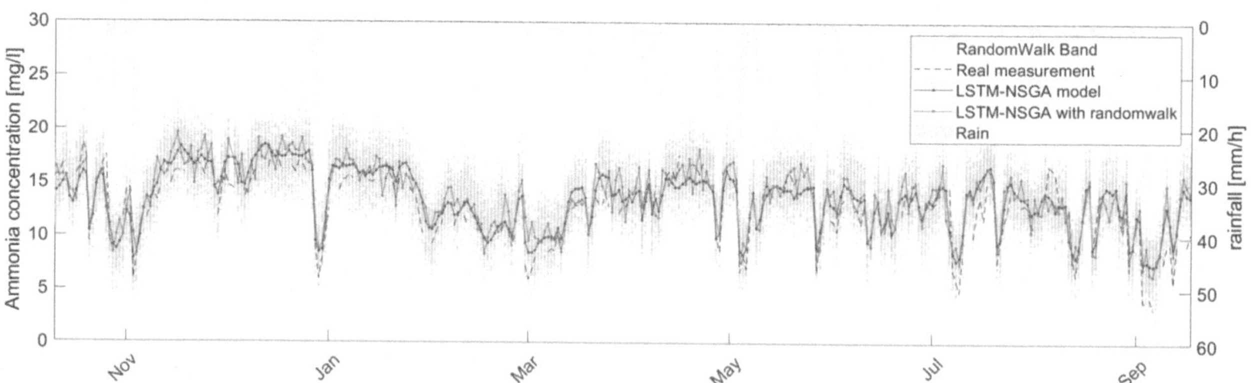

Figure 10 | Ammonia concentration reconstruction with random walk process: the grey band represents the confidence interval generated using 200 Monte Carlo simulations.

The proposed IG model was validated and tested on an actual case study, and it achieves excellent performance in both MAPE and time series distribution criteria. With the addition of a random walk process, it is able to generate a time series with a probability distribution that is even more similar to the observed reality, and thus gives better influent description for WRRF design and operation decision-making. It also allows for gap filling of incomplete databases, for instance for nitrogen species that are sparsely measured at carbon-only removing plants as studied in this work.

ACKNOWLEDGEMENTS

Sincere thanks are extended to SUEZ SGAC and NSERC (Natural Sciences and Engineering Research Council of Canada) for funding the research project. Québec City is thanked for its contribution with all relevant data. Peter Vanrolleghem holds the Canada Research Chair on Water Quality Modelling.

DATA AVAILABILITY STATEMENT

All relevant data are included in the paper or its Supplementary Information.

REFERENCES

Ahnert, M., Marx, C., Krebs, P. & Kuehn, V. 2016 A black-box model for generation of site-specific WWTP influent quality data based on plant routine data. *Water Sci. Technol.* **74**, 2978–2986.

Alferes, J., Tik, S., Copp, J. & Vanrolleghem, P. A. 2013 Advanced monitoring of water systems using in situ measurement stations: data validation and fault detection. *Water Sci. Technol.* **68**, 1022–1030.

Aminabad, M., Maleki, A., Hadi, M. & Shahmoradi, B. 2013 Application of Artificial Neural Network (ANN) for the prediction of water treatment plant influent characteristics. *J. Adv. Environ. Health Res.* **1**, 89–100.

Borzooei, S., Teegavarapu, R., Abolfathi, S., Amerlinck, Y., Nopens, I. & Zanetti, M. C. 2019 Data mining application in assessment of weather-based influent scenarios for a WWTP: getting the most out of plant historical data. *Water Air Soil Pollut.* **230**, 5.

Chiandussi, G., Codegone, M., Ferrero, S. & Varesio, F. E. 2012 Comparison of multi-objective optimization methodologies for engineering applications. *Comput. Math. Appl.* **63**, 912–942.

Corominas, L., Garrido-Baserba, M., Villez, K., Olsson, G., Cortés, U. & Poch, M. 2018 Transforming data into knowledge for improved wastewater treatment operation: a critical review of techniques. *Environ. Modell. Software* **106**, 89–103.

Deb, K., Pratap, A., Agarwal, S. & Meyarivan, T. 2002 A fast and elitist multiobjective genetic algorithm: NSGA-II. *IEEE Trans. Evol. Comput.* **6**, 182–197.

El-Din, A. G. & Smith, D. W. 2002 A neural network model to predict the wastewater inflow incorporating rainfall events. *Water Res.* **36**, 1115–1126.

El-Din, A. G., Smith, D. W. & El-Din, M. G. 2004 Application of artificial neural networks in wastewater treatment. *J. Environ. Eng. Sci.* **3**, S81–S95.

Ercan, M. B. & Goodall, J. L. 2016 Design and implementation of a general software library for using NSGA-II with SWAT for multi-objective model calibration. *Environ. Modell. Software* **84**, 112–120.

Flores-Alsina, X., Saagi, R., Lindblom, E., Thirsing, C., Thornberg, D., Gernaey, K. V. & Jeppsson, U. 2014 Calibration and validation of a phenomenological influent pollutant disturbance scenario generator using full-scale data. *Water Res.* **51**, 172–185.

Fonseca, C. & Fleming, P. 1999 Genetic algorithms for multiobjective optimization: formulation discussion and generalization. In: *Proceedings of the 5th International Conference on Genetic Algorithms*, San Francisco, United States, pp. 416–423.

Gernaey, K. V., Flores-Alsina, X., Rosen, C., Benedetti, L. & Jeppsson, U. 2011 Dynamic influent pollutant disturbance scenario generation using a phenomenological modelling approach. *Environ. Modell. Software* **26**, 1255–1267.

Goldberg, D. E. 1989 *Genetic Algorithms in Search, Optimization and Machine Learning*, 1st edn. Addison-Wesley Longman Publishing Co., Inc., MA, United States.

Graves, A. 2012 Supervised sequence labelling with recurrent neural networks, Vol. 385. In: (Graves, A. (ed)) *Studies in Computational Intelligence*. Springer Berlin Heidelberg, Berlin, Heidelberg.

Hassanat, A., Almohammadi, K., Alkafaween, E., Abunawas, E., Hammouri, A. & Prasath, V. B. S. 2019 Choosing mutation and crossover ratios for genetic algorithms – a review with a new dynamic approach. *Information* **10**, 390–402.

Hauduc, H., Neumann, M. B., Muschalla, D., Gamerith, V., Gillot, S. & Vanrolleghem, P. A. 2015 Efficiency criteria for environmental model quality assessment: a review and its application to wastewater treatment. *Environ. Modell. Software* **68**, 196–204.

Hochreiter, S. 1998 The vanishing gradient problem during learning recurrent neural nets and problem solutions. *Int. J. Uncertainty Fuzziness Knowlege Based Syst.* **6**, 107–116.

Hochreiter, S. & Schmidhuber, J. 1997 Long short-term memory. *Neural Comput.* **9**, 1735–1780.

Holland, J. H. 1992 *Adaptation in Natural and Artificial Systems: An Introductory Analysis with Applications to Biology, Control, and Artificial Intelligence*. MIT Press, Cambridge, Mass

Konak, A., Coit, D. W. & Smith, A. E. 2006 Multi-objective optimization using genetic algorithms: a tutorial. *Reliab. Eng. Syst. Saf.* **91**, 992–1007.

Kullback, S. & Leibler, R. A. 1951 On information and sufficiency. *Ann. Math. Stat.* **22**, 79–86.

Langeveld, J., Van Daal, P., Schilperoort, R., Nopens, I., Flameling, T. & Weijers, S. 2017 Empirical sewer water quality model for generating influent data for WWTP modelling. *Water* **9**, 1–18.

Levenberg, K. 1944 A method for the solution of certain non-linear problems in least squares SQUARES. *Quarterly of Applied Mathematics* **2** (2), 164–168. DOI:10.1090/QAM/10666.

Li, F., Amaral, A. & Vanrolleghem, P. A. 2020 An essential tool for WRRF modelling: a realistic and complete influent generator for flow rate and water quality based on machine learning. In: *93rd Water Environment Federation Technical Exhibition and Conference 2020, WEFTEC 2020.* Water Environment Federation, pp. 303–309.

Ma, S., Zeng, S., Dong, X., Chen, J. & Olsson, G. 2014 Short-term prediction of influent flow rate and ammonia concentration in municipal wastewater treatment plants. *Front. Environ. Sci. Eng.* **8**, 128–136.

Martin, C. & Vanrolleghem, P. A. 2014 Analysing, completing, and generating influent data for WWTP modelling: a critical review. *Environ. Modell. Software* **60**, 188–201.

Miller, B. L. & Goldberg, D. E. 1996 Genetic algorithms, selection schemes, and the varying effects of noise. *Evol. Comput.* **4**, 113–131.

Newhart, K. B., Marks, C. A., Rauch-Williams, T., Cath, T. Y. & Hering, A. S. 2020 Hybrid statistical-machine learning ammonia forecasting in continuous activated sludge treatment for improved process control. *J. Water Process Eng.* **37**, 101389.

Raman, H. & Sunilkumar, N. 1995 Multivariate modelling of water resources time series using artificial neural networks. *Hydrol. Sci. J.* **40**, 145–163.

Reichert, P. & Schuwirth, N. 2012 Linking statistical bias description to multiobjective model calibration. *Water Resour. Res.* **48**, 1–20.

Rojas, R. 1996 Chapter 17 genetic algorithms. In: (Rojas, R. (ed)) *Neural Networks: A Systematic Introduction.* Springer-Verlag, Berlin, Heidelberg, pp. 429–450.

Schlichthärle, D. 2011 *Analog Filters – Digital Filters: Basics and Design.* Springer Berlin Heidelberg, Berlin, Heidelberg.

Shokry, A., Vicente, P., Escudero, G., Pérez-Moya, M., Graells, M. & Espuña, A. 2018 Data-driven soft-sensors for online monitoring of batch processes with different initial conditions. *Comput. Chem. Eng.* **118**, 159–179.

Srinivas, M. & Patnaik, L. M. 1994 Adaptive probabilities of crossover and mutation in genetic algorithms. *IEEE Trans. Syst. Man Cybern.* **24**, 656–667.

Talebizadeh, M., Belia, E. & Vanrolleghem, P. A. 2015 Probabilistic design of wastewater treatment plants. In: *Proceedings 88th Annual WEF Technical Exhibition and Conference, WEFTEC 2015*, Chicago, IL, USA.

Tik, S. & Vanrolleghem, P. A. 2017 Chemically enhancing primary clarifiers: model-based development of a dosing controller and full-scale implementation. *Water Sci. Technol.* **75**, 1185–1193.

Villez, K., Del Giudice, D., Neumann, M. B. & Rieckermann, J. 2020 Accounting for erroneous model structures in biokinetic process models. *Reliab. Eng. Syst. Saf.* **203**, 107075.

Wang, Q., Wang, L., Huang, W., Wang, Z., Liu, S. & Savić, D. A. 2019 Parameterization of NSGA-II for the optimal design of water distribution systems. *Water* **11**, 971.

Yusoff, Y., Ngadiman, M. S. & Zain, A. M. 2011 Overview of NSGA-II for optimizing machining process parameters. *Procedia Eng.* **15**, 3978–3983.

Zhang, Q., Li, Z., Snowling, S., Siam, A. & El-Dakhakhni, W. 2019 Predictive models for wastewater flow forecasting based on time series analysis and artificial neural network. *Water Sci. Technol.* **80**, 243–253.

First received 30 September 2021; accepted in revised form 27 January 2022. Available online 8 February 2022

doi: 10.2166/wst.2022.115

Hybrid modelling of water resource recovery facilities: status and opportunities

Mariane Yvonne Schneider [a,*], Ward Quaghebeur [b,c,d], Sina Borzooei [b,c], Andreas Froemelt [e], Feiyi Li [f], Ramesh Saagi [g], Matthew J. Wade [h], Jun-Jie Zhu [i] and Elena Torfs [b,c]

[a] Next Generation Artificial Intelligence Research Center & School of Information Science and Technology, The University of Tokyo, 7-3-1 Hongo, Bunkyo-ku, Tokyo 113-8656, Japan
[b] Centre for Advanced Process Technology for Urban Resource recovery (CAPTURE), Frieda Saeysstraat 1, Gent 9000, Belgium
[c] BIOMATH, Department of Data Analysis and Mathematical Modelling, Ghent University, Coupure Links 653, Ghent 9000, Belgium
[d] KERMIT, Department of Data Analysis and Mathematical Modelling, Ghent University, Coupure Links 653, Ghent 9000, Belgium
[e] Eawag, Swiss Federal Institute of Aquatic Science and Technology, Dübendorf 8600, Switzerland
[f] modelEAU, CentrEau, Département de génie civil et de génie des eaux, Pavillon Adrien-Pouliot, Université Laval, Quebec City, Canada
[g] Division of Industrial Electrical Engineering and Automation (IEA), Department of Biomedical Engineering, Lund University, P.O. Box 118, Lund SE-22100, Sweden
[h] School of Engineering, Newcastle University, Newcastle-upon-Tyne NE1 7RU, UK
[i] Department of Civil and Environmental Engineering and Andlinger Center for Energy and the Environment, Princeton University, Princeton, NJ 08544, USA
*Corresponding author. E-mail: myschneider@isi.imi.i.u-tokyo.ac.jp

MYS, 0000-0003-3397-2773; WQ, 0000-0002-6162-2124; SB, 0000-0002-0694-3064; AF, 0000-0001-9388-7816; FL, 0000-0003-4278-730X; RS, 0000-0003-4373-2562; MJW, 0000-0001-9824-7121; J-JZ, 0000-0002-7546-2870; ET, 0000-0002-5629-6950

ABSTRACT

Mathematical modelling is an indispensable tool to support water resource recovery facility (WRRF) operators and engineers with the ambition of creating a truly circular economy and assuring a sustainable future. Despite the successful application of mechanistic models in the water sector, they show some important limitations and do not fully profit from the increasing digitalisation of systems and processes. Recent advances in data-driven methods have provided options for harnessing the power of Industry 4.0, but they are often limited by the lack of interpretability and extrapolation capabilities. Hybrid modelling (HM) combines these two modelling paradigms and aims to leverage both the rapidly increasing volumes of data collected, as well as the continued pursuit of greater process understanding. Despite the potential of HM in a sector that is undergoing a significant digital and cultural transformation, the application of hybrid models remains vague. This article presents an overview of HM methodologies applied to WRRFs and aims to stimulate the wider adoption and development of HM. We also highlight challenges and research needs for HM design and architecture, good modelling practice, data assurance, and software compatibility. HM is a paradigm for WRRF modelling to transition towards a more resource-efficient, resilient, and sustainable future.

Key words: data-driven model, hybrid model, mechanistic model, process control, urban water management, wastewater

HIGHLIGHTS

- HM combines mechanistic and data-driven models.
- WRRFs are too sensitive to rely only on data-driven techniques; HM is an alternative.
- Hybrid models have a high predictive power at low computational cost, proving them useful for online optimisation and control.
- Major challenges for more widespread implementation of HM are discussed.
- HM supports the transition towards a digital and resource-efficient water sector.

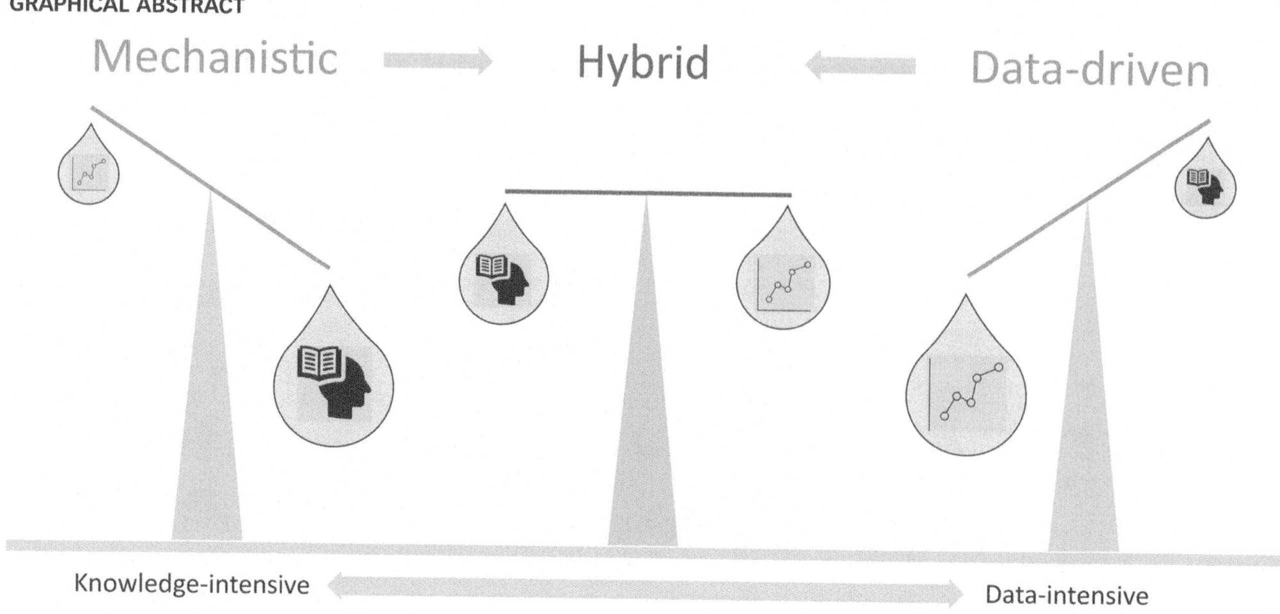

ABBREVIATIONS

ADM	anaerobic digestion model
AI	artificial intelligence
ANN	artificial neural networks
Anammox	anaerobic ammonium oxidation
AOB	ammonia-oxidising bacteria
API	application programming interface
ARIMA	autoregressive integrated moving average
ASM	activated sludge model
BSM	benchmark simulation model
CFD	computational fluid dynamics
CSTR	continuously stirred tank reactor
DAE	differential-algebraic equation
GLUE	generalised likelihood uncertainty estimation
GMP	good modelling practice
HM	hybrid modelling
kNN	k-nearest neighbours
LSTM	long short-term memory
ML	machine learning
MLP	multilayer perceptrons
MPC	model predictive control
NLP	natural language processing
NOB	nitrite-oxidising bacteria
ODE	ordinary differential equation
PCA	principal component analysis
PDE	partial differential equation
PLS	partial least square
RBF	radial basis functions
SVM	support vector machine
WRRF	water resources recovery facility
WRRmod	water resource recovery modelling
WTP	water treatment plant

1. INTRODUCTION AND MOTIVATION

Urban water management has a critical function for maintaining human and environmental health by ensuring potable water quality and treated wastewater. This goal can be reached by operating within defined safety margins or through adequate monitoring, control, and interventions, or a combination of the two. Conservative design and operation, employing high safety margins, typically lead to oversized plants and energy-intensive processes, resulting in increased costs and resource consumption. Such approaches can improve reliability or resilience for future changes. However, urban water management increasingly and critically faces systemic challenges that cannot be ignored, such as resource shortages (e.g., clean water, energy, nutrients), population growth, and climate change. Therefore, new approaches that support activities to address these challenges and improve the sustainability and resilience of urban water management systems (Butler *et al.* 2017) are critical research needs. Here, we focus on the considerable utility of mathematical modelling, specifically the potential of hybrid models.

Water resources recovery facilities (WRRFs) are an important part of urban water management and have complex, partially observable processes that require monitoring and control to ensure stable, safe, and appropriate operation (Olsson & Newell 1999). Modelling biological, chemical, and physical processes is useful to acquire process understanding, simulate and test control strategies, predict future behaviour under changing conditions (Gernaey *et al.* 2004), or model-predictive control (Lund *et al.* 2018). In recent decades, mechanistic approaches to model wastewater treatment processes have been favoured by engineers due to the first-principles approach well established across many industries (Henze *et al.* 2000; Newhart *et al.* 2019). These mechanistic models are based on process knowledge. These models usually simplify the complex processes they represent, such as aeration and mixing in activated sludge plants or aggregation of particulates. A concrete example is the diversity of nitrifying bacteria known to function in these systems (e.g., Daims 2014), which are typically summarised in models as nitrifying bacteria or differentiated into a few species, i.e., ammonia-oxidising bacteria (AOB) and nitrite-oxidising bacteria (NOB). Furthermore, and especially with greater model complexity, parameterising, calibrating, and validating mechanistic models usually require laborious experiments that may not always be possible due to economic, time, or measurability constraints (Vanrolleghem *et al.* 2005). While activated sludge models (ASMs) (Henze *et al.* 2000) are widely accepted in both research and practice, modelling several underlying processes (e.g., nitrous oxide production, phosphorus removal processes) remains incomplete (Regmi *et al.* 2019). Similarly, a standard and widely accepted modelling framework for several novel treatment processes (e.g., anaerobic ammonium oxidation (Anammox) processes (Baeten *et al.* 2019), membrane treatment (Mannina *et al.* 2011)) is not yet available. Such modelling bottlenecks can result in hindered innovation and rapid system implementation.

While data-driven methods have always been part of a modeller's toolbox, they are currently receiving greater interest, motivated by cheaper and smarter sensor technologies that enable ubiquitous data collection as well as online and real-time monitoring (Vanrolleghem & Lee 2003; Corominas *et al.* 2018). This has resulted in an increasing amount of high-resolution data, which together with available computational storage, cheap, edge computation, modern database technology, data processing (e.g., Hadoop (Apache Software Foundation n.d.)), and transmission capacity maximise the extraction of useful information available to plant operators (Wade 2004; Samuelsson *et al.* 2021). The need for novel and improved methods in combination with greater opportunities for data-driven approaches has induced a significant body of literature discussing the use of such methods in the water processing industry (Wade *et al.* 2021). These methods are particularly successful on problems involving large and high-quality data sets (Dobbelaere *et al.* 2021). However, data quality might be a barrier, especially in the wastewater treatment sector, where sensors are exposed to harsh conditions (e.g., Cecconi *et al.* 2020; Samuelsson *et al.* 2021). An additional issue of the data-driven approach is a lack of mechanistic-based interpretability of the result and limitations in extrapolation power (Newhart *et al.* 2019).

In this article, one solution to bring forth the advantages of both mechanistic and data-driven models and minimise their deficits is hybrid modelling (HM). HM is a combination of mechanistic and data-driven modelling. The mechanistic models ensure the preservation of the process knowledge and extrapolation capabilities of the model. The data-driven models offer improved predictive capabilities that can alleviate the weaknesses in mechanistic models, for example, by learning unknown, hidden relationships.

The water sector's transition towards a circular economy has created a shift in modelling objectives. For example, more case-specific quality targets for different water reuse purposes might be defined, which require tailor-made monitoring and modelling (Reynaert *et al.* 2021). Simultaneously, the digitalisation of the water sector, characterised by developments in sensor and IT technology, is creating vast amounts of data and, as such, the need for increased computational capacity and storage. This provides both more input data for modelling but also allows models to be evaluated at finer timescales, e.g., real-time or near real-time, which is particularly beneficial for improving process automation and control. These combined benefits create an environment for hybrid models to be used to their full potential by simultaneously harnessing and integrating the available data, computational power, and enhanced information on the studied processes. As such, the application of HM in the domain of WRRFs has the potential to foster automation (Rodriguez-Roda *et al.* 2002), increase efficiency, and increase the predictive power of models (von Stosch *et al.* 2014). Furthermore, we argue that HM helps keep up with new technologies and processes by being, for example, a preceding step in developing a mechanistic model prior to acquiring sufficient process understanding. However, uptake of hybrid models in the WRRF modelling community is hampered by the absence of expertise, trust in the novel (i.e. data-driven) approaches, and documented methods for the development and robust implementation of HM.

To better coordinate advances of HM in the WRRF sector, a seminar on HM was organized at the specialist WRR modelling (WRRmod) 2021 conference in Arosa, Switzerland. As an output from this, we provide a comprehensive overview of HM philosophy and architecture. We give examples of successful implementations in the WRRF sector and discuss challenges, opportunities, and development needs for hybrid models to develop and be integrated by the industry as a powerful framework for WRRF applications.

In this article, we use the terminology mechanistic, data-driven, and hybrid modelling to refer to the different modelling paradigms. This deviates from previous work, e.g., von Stosch *et al.* (2014), who refer to mechanistic models as parametric since their parameters have a physical meaning. This is not the case in data-driven models, hence their designation as non-parametric. Hybrid models, falling somewhere on the spectrum between the two approaches, are referred to as semi-parametric. However, this usage can be confusing. In statistics, parametric is defined as *having a finite number of parameters*. Thus, both mechanistic models and certain types of data-driven models (e.g., artificial neural networks (ANNs)) fall within this definition. In contrast, other types of data-driven models (e.g., k-nearest neighbours (kNNs) or support vector machine (SVM)) are considered non-parametric. To limit confusion, we do not use the parametric terminology in this manuscript but refer to von Stosch *et al.* (2014) as an appropriate review of HM on the broader process industry.

2. BACKGROUND ON HM

Following Schubert *et al.* (1994), we define two types of knowledge: mechanistic knowledge and knowledge learned from any patterns encapsulated within data. To enable a deep discussion of the intricacies of hybrid models, it is necessary first to discuss two modelling paradigms based on these types of knowledge: mechanistic and data-driven models.

Mechanistic models are models based on foundational knowledge. Mechanistic knowledge provides a representative description of a system or process derived from a fundamental understanding of its underlying mechanisms based on physical or chemical laws or causal relations. This type of knowledge is then encapsulated in a set of mathematical functions, whose behaviour can be described by one or more mathematical equations. We suggest that mechanistic models comprise elements of first-principles (i.e. the foundational understanding of a process or system) and phenomenological (i.e. describing empirical relationships of process/system components) models. First-principles models are completely based on natural laws without relying on data, whereas phenomenological or empirical models are structured according to domain knowledge, but data determine their parameters. Examples of the latter are Monod kinetics in ASM (Henze *et al.* 2000) or settling velocity functions in settler models (Takács *et al.* 1991).

Simple algebraic equations can often adequately describe the system. However, for more complex systems, the underlying dynamics are often explicitly described using ordinary differential equations (ODEs), differential-algebraic equations (DAEs) or partial differential equations (PDEs). In addition, there is a wide range of alternatives for equation-based models, including agent-based models, population balance models, and equation-free modelling (DeAngelis & Yurek 2015). However, these are not commonly applied as models for water and resource recovery applications, so we focus the discussion on models based on algebraic and differential equations.

However, process knowledge is often limited and incomplete, where the model does not consider latent or unobservable mechanisms. This typically results in only a partial description of the process or system, which may have value but often is not robust or resilient to significant changes.

Data-driven models derive knowledge from patterns encapsulated by measured data and metadata. In contrast to mechanistic models, no domain knowledge is assumed. It is important to distinguish between data-driven models and empirically derived phenomenological models. The former does not assume any structure in the data; instead, it learns the structure from the data itself. At the same time, the latter defines a structure from current domain knowledge, determining the values of its parameters from data. This inclusion of domain knowledge classifies them as mechanistic.

Examples of frequently used data-driven methods are Gaussian processes, polynomials, multivariate adaptive regression spline, partial least squares (PLS), principal component analysis (PCA), SVM, decision tree-based methods, and ANNs. In the context of machine learning (ML), data-driven methods and models can be categorised based on their learning algorithms, i.e. supervised, unsupervised, and reinforcement learning. For methods based on supervised learning, including classification and regression algorithms, the mapping function is learned from input and output data and their associations (Razavi 2021). For unsupervised learning problems, such as clustering and association, a model learns from unlabelled input data and discovers common aspects in unknown patterns (Abrahart *et al.* 2008; Zhu *et al.* 2018; Borzooei *et al.* 2020). Reinforcement learning (Sutton & Barto 2018), where the model or agent learns using trial and error or through a reward-punishment process, has found favour more recently as a more knowledge-based and adaptive method for plant control (Hernández-del-Olmo *et al.* 2012; Pang *et al.* 2019; Lazăr *et al.* 2020). We recommend, for example, Mitchell (1997) for a more comprehensive discussion of data-driven methods.

As no domain knowledge is included, the success of the data-driven approach depends heavily on the quality and quantity of the data and relies on an appropriate choice of method for extraction. Moreover, predictions are not designed to obtain a causal relation. Therefore, the interpretability of data-driven models is a critical bottleneck.

HM emerged to combine the use of the two aforementioned modelling paradigms (Figure 1). The first research to combine these paradigms can be traced back to the early 1990s (compare section 3, text mining). Early hybrid models consisted of a combination of either ANNs or radial basis functions (RBFs) with mathematical expressions describing mass and energy balances, respectively data-driven and mechanistic components (Johansen & Foss 1992; Kramer *et al.* 1992; Psichogios & Ungar 1992; Su *et al.* 1993). As hybrid models were simultaneously developed in different research fields, the terminology used differed; The de facto name hybrid model was initially used in the field of chemical engineering. Other terms for HM include but are not limited to: (i) Hybrid semi-parametric modelling (e.g., von Stosch *et al.* 2014), mainly to emphasise the use of both parametric (i.e. mechanistic) and non-parametric (i.e. data-driven) models; (ii) Knowledge-based HM, highlighting the fact

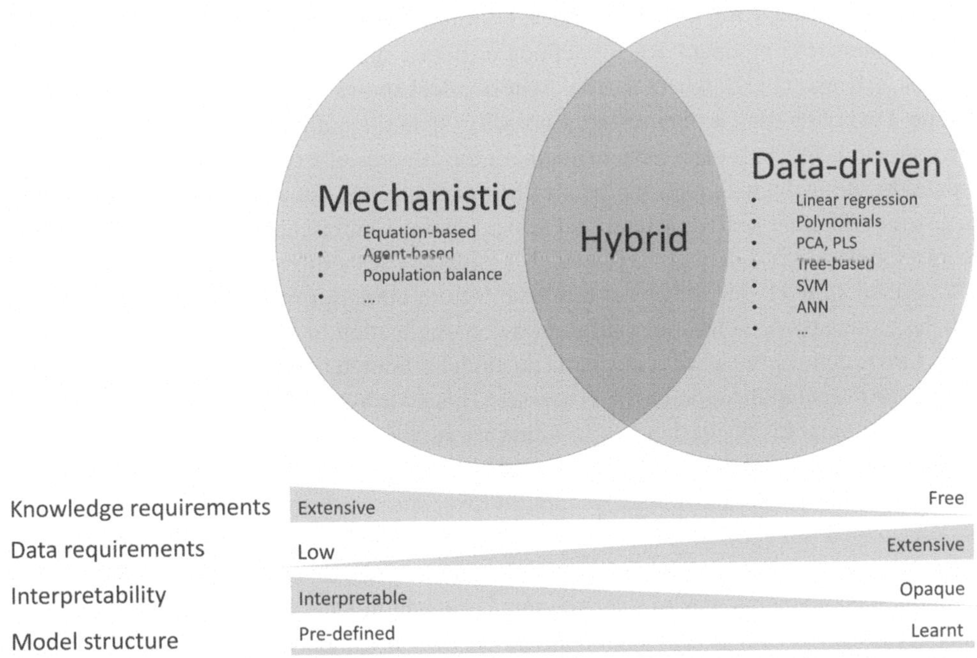

Figure 1 | Venn diagram of hybrid modelling, schematically showing the requirements of expert knowledge and data, the interpretability, and the model structure.

that all knowledge available (data, information, and domain expert knowledge) is incorporated into the model; (iii) Integrated neural network is used for an approach similar to HM (Su *et al.* 1993). (iv) Grey-box modelling, where the terminology arose, reflects a mixture of the black- and white-box concepts (Tulleken 1993; Van Can *et al.* 1996). Black comes from the black-box model to indicate that the exact relation of the mapping between input and output is often difficult to explain and does not necessarily have physical interpretability. In contrast, a white-box model corresponds to a mechanistic model, where the relationships themselves are defined by modelling experts and therefore assumed to be known. The black-box model is often associated with the data-driven approach; however, this use is controversial, both in the word's negative connotations and the confusing terminology, e.g., with linear regression or decision trees, the mapping is transparent, apropos our suggested terminology. Therefore, we refrain from using this terminology in this manuscript.

The original intention of hybrid models was to impute information absent from mechanistic models using available (i.e. measurable) data (Psichogios & Ungar 1992). The domain knowledge used to construct a mechanistic model is often incomplete and subject to several simplifying assumptions. The inverse motivation, i.e. to integrate mechanistic components into data-driven models, is for example, present in climate research (Karpatne *et al.* 2017a). Here, techniques such as theory-guided data science and physics-guided or physics-informed neural networks are built on high data volumes and subjected to certain physical constraints (Karpatne *et al.* 2017b; Raissi *et al.* 2017, 2019; Luo & Bao 2018).

Central to a hybrid model is using a data-driven component that does not assume a structure. Solely using a phenomenological or empirical component within a mechanistic model does not make it a hybrid model, as this component still assumes a structure derived from domain knowledge (e.g., the usage of an empirical Monod function in the ASM model (Henze *et al.* 2000)). Different architectures have been investigated for combining mechanistic and data-driven components with hybrid models. We distinguish between a strict and a loose definition for HM. The strict definition requires that both sub-models must be present in the final hybrid model. Two architectures, serial and parallel (von Stosch *et al.* 2014), fit this description. When loosening this definition by requiring both paradigms to be used during model development, other paradigms such as surrogate modelling (Razavi *et al.* 2012), also known as emulation, fit under the HM umbrella. The latter is specifically useful when the function of existing models is constrained by the computational ability or observability of the outputs (Oyebamiji *et al.* 2019). Our recommendations apply to the loose definition as they are addressed towards any expert using both approaches. Figure 2 schematically displays the three model architectures.

The first architecture adhering to the strict definition, a **serial** hybrid model, is arguably the most used. In this architecture, the output of one model is used as input to the other. Usually, the output of the data-driven model is used as input of the mechanistic model (e.g., Côté *et al.* 1995). The mechanistic component incorporates the known dynamics, usually chemical and physical phenomena, such as mass and energy balances. The data-driven component outputs dynamics that are not explicitly modelled by the mechanistic component, e.g., ill-defined or poorly understood reaction kinetics. An early example is a work done by Psichogios & Ungar (1992), where a serial hybrid model showed higher accuracy than either a mechanistic or data-driven model alone. Less often used is the inverse approach, where the output of a mechanistic model is embedded into or used by a data-driven model. This approach aims to augment the features of a data-driven model with domain knowledge, directly feeding the data-driven model with relevant data transformations. A pertinent example of this is the representation of inter-connected processes at different spatial or temporal scales, where a multiscale approach utilises both mechanistic representation for dynamic simulation and data-driven methods for higher=scale inference of process behaviour (Li *et al.* 2019; Hannaford *et al.* 2021). Tsen *et al.* (1996) show that this architecture outperforms other architectures and the mechanistic model itself. Nevertheless, this approach has not yet found much application in practice.

The **parallel** architecture is compelling when a mechanistic model is limited in describing the dynamics, e.g., due to incomplete domain knowledge or oversimplification of the processes. This contrasts with the serial hybrid model, which excels in situations where a specific subprocess is not defined. In a first variant, a cooperative parallel hybrid model, the data-driven model is trained to learn the mismatch between the mechanistic model and historical data, i.e. the residual error. Then both results are fused, usually by addition. A recent improvement in this approach is to learn the dynamics' residuals rather than the state's residuals (Quaghebeur *et al.* 2022). In a second variant, a competitive parallel hybrid model, the mechanistic and data-driven models are trained to make the same prediction. In other words, this technique can be considered an ensemble approach, in which different components seek to solve the same problem independently to improve the ensemble model performance (Hu *et al.* 2011; Abba *et al.* 2020). These predictions are then weighted and combined in a final output (Dors *et al.* 1995; Peres *et al.* 2001; Galvanauskas *et al.* 2004; Ghosh *et al.* 2019).

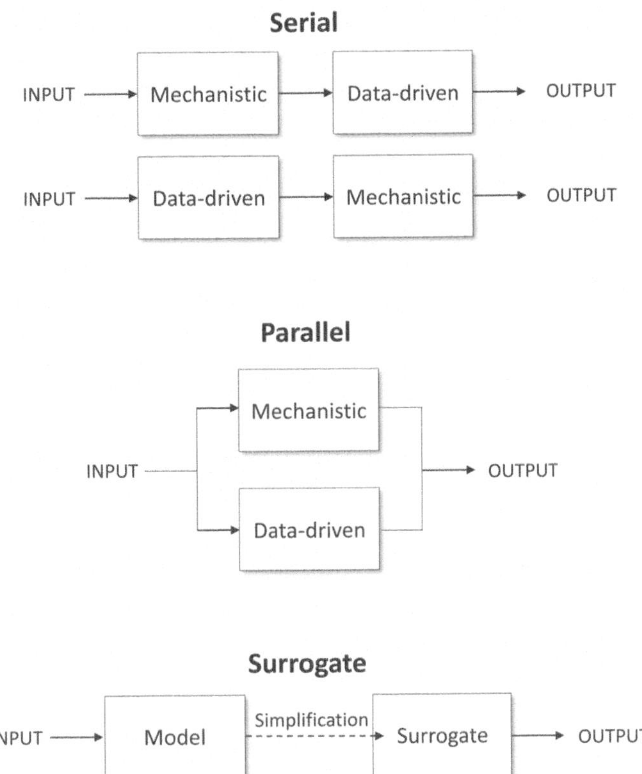

Figure 2 | Schematics of the three model architectures used in HM.

When we apply the loose definition of hybrid models, it typically implies the use of **surrogate models**, also known as meta-models (Jin *et al.* 2001), substitute models (Romijn *et al.* 2008), emulators (Conti & Hagan 2010; Mahmoodian *et al.* 2018), or response surface models (Wang & Georgakis 2019). These models are trained on the output of a mechanistic model to create a computationally less demanding model for rapid or large-scale simulations, such as required by, for example, real-time model predictive control, uncertainty analysis, or optimisation problems.

After defining the model architecture, whether serial, parallel, or surrogate, the parameters of the model need to be estimated from data, a process also known as parameter identification. Mechanistic modellers refer to this process as calibration, while data-driven modellers usually prefer the term training. The identification procedure usually aims to minimise the error between data and model output, i.e. by optimising the model fit. In both serial and parallel hybrid models, the parameter values of both components can be identified either separately or in conjunction.

In the separate approach, the parameters of one of the models are identified before identifying the other model. Usually, the mechanistic model is identified first, with the data-driven model identified afterwards. If needed, this procedure can be iterated several times. However, convergence is not guaranteed nor stable. Azarpour *et al.* (2017) suggest good modelling practices for hybrid model identification following the separate approach.

In the conjunction approach, the parameter values of both models are identified simultaneously. For example, the hybrid model is trained by backpropagating errors through the data-driven and mechanistic models (Psichogios & Ungar 1992). This involves tracking the gradients of both components with either nonlinear optimisation algorithms, such as sequential quadratic programming. Recently, methods such as automatic differentiation or adjoint sensitivity calculation have been adapted to the context of differential equations (Chen *et al.* 2018), making them applicable in the context of HM (Quaghebeur *et al.* 2021).

3. HM IN THE WATER INDUSTRY

A deep, comprehensive text mining analysis was conducted to inventory HM research publications related to the field of wastewater treatment. First, we used general terms (e.g., 'wastewater', 'sewage', 'activated sludge', and 'anaerobic digestion') to retrieve more than 320,000 wastewater related publication records (such as title, abstract, author keywords, etc.; updated on January 1, 2022) from the Web of Science (WOS 2021). We adopted a series of natural language processing (NLP) pre-treatment steps to inventory the data for year, title, abstract, author keywords, and correspondence (Zhu *et al.* 2021). The NLP-based data preprocessing included N-grams generation (N=1–5, or tokenisation to one, two, three, four, and five adjacent words), lower-case/punctuation/stop-words treatment, and stemming (convert a word or term to its root form; Manning *et al.* (2008)). Meanwhile, we prepared four lists of terminologies (L) that were used as surrogates to represent (L1) wastewater treatment (>90 terms; e.g., wastewater, sewage, activated sludge, etc.), (L2) data-driven modelling only (>180 terms; e.g., regression, ML, ANN, etc.), (L3) mechanistic modelling only (>70 terms; e.g., mechanistic model, first-principles model, ASM, etc.), and (L4) hybrid modelling (e.g., gray-/grey-box, etc.). Two combinations ('L1 AND L2 AND L3' and 'L1 AND L4') of the key lists were then used to identify relevant papers from the initial 320,000 raw data. Specifically, the data of year and correspondence were used to identify variations of temporal (1990–2022) and spatial distribution. The spatial distribution analysis was performed based on the origin of the corresponding author(s) and countries or regions (Zhu *et al.* 2021). The dictionary of preprocessed textual corpora was used to identify hybrid model-based (the two combinations described above), data-driven model-based (e.g., supervised or/and unsupervised learning), and mechanistic model-based (e.g., variations of ASM and the anaerobic digestion model No. 1 (ADM)) publications based on their corresponding lists of terminology-surrogates. Furthermore, 11 representative terminologies (and their synonyms) were selected and used to showcase their research popularity in the HM publications. It is worth noting that model predictive control (MPC) studies were also identified from the raw wastewater database; not all MPC studies used a HM framework, but MPC can be a typical HM application in wastewater management.

In the 320,000 wastewater-related publications, more than 9,800 papers studied data-driven relevant methods and more than 3,900 papers focused on mechanistic models; their intersection, hybrid models, accounts for 534 publications. The number of relevant papers for HM increased steadily over the years from less than 10 in the early 1990s to more than 50 in 2021 (Figure 3(a)). Among the 534 papers, the majority of the HM research papers (506 or 94.8%) are identified based on the combination of data-driven and mechanistic terminologies. Specific HM terminologies, such as grey-box modelling, were used to retrieve 24 papers (4.5%) (Figure 3(c)). From the spatial perspective, the relevant papers were authored from 56 countries or regions; and the annual spatial popularity increased from less than five in the early 1990s to more than 20 countries or regions by 2021 (Figure 3(b)). Specifically, most of the papers (corresponding) were authored from Europe (224 or 41.9%), Asia (212 or 39.7%), and North America (56 or 10.5%) (Figure 3(d)). From the data-driven methods perspective, 324 papers studied supervised learning method(s), whereas only 32 papers explicitly used unsupervised learning method(s), and only six studies were related to reinforcement learning ((Figure 3(e)). The remaining papers did not specify the data-driven methods, including grey-box modelling and other miscellaneous terms, such as machine learning, big data, artificial intelligence, etc. A similar analysis applied to mechanistic models shows that 166 papers focused on the ASMs or ADM, 111 papers studied benchmark simulation models (BSMs) or related, and the remaining 257 papers did not specify any mechanistic models in their abstracts. A closer analysis of the 257 'non-explicit' papers shows that 57 papers studied Monod kinetics, and 12 described relevant process simulation software. As previously mentioned, MPC can be one HM application, and 276 studies have investigated MPC applied to wastewater management. In addition, the specific terminology inventory also shows that more studies used ML (250) than classical statistical methods (126) (e.g., multiple linear regression, autoregressive integrated moving average (ARIMA), PLS); among the ML methods, neural networks (146) are the dominant type (Figure 3(g)).

The text mining analyses reveal several valuable observations. First, the term 'grey-box modelling' (28) was not frequently used, while the majority of HM studies were identified based on a combination of using both data-driven and mechanistic models (Figure 3(g)). Second, a rapidly increasing number of papers based on both temporal and spatial variations started about 10–20 years ago, matching the period when the data-driven methods gained increasing popularity due to the recent advances in ML. Third, supervised learning is the dominant data-driven application, probably because data obtained from this approach can be more readily used for subsequent process management, while unsupervised learning is typically used to cluster different scenarios, so its applications are more often limited to off-site scenario analyses (Zhu *et al.* 2015; Borzooei *et al.* 2019). Though the number of papers is low, reinforcement learning or active learning may potentially improve process control when labelled data is not largely available (Russo *et al.* 2020). Furthermore, it seems that there are two ways to implement and test HM; (i) adaptation of ASM/ADM to hybridise with data-driven methods when data is not a limiting factor; (ii) testing process optimisation based on benchmarking models, such as BSM, with simulated data.

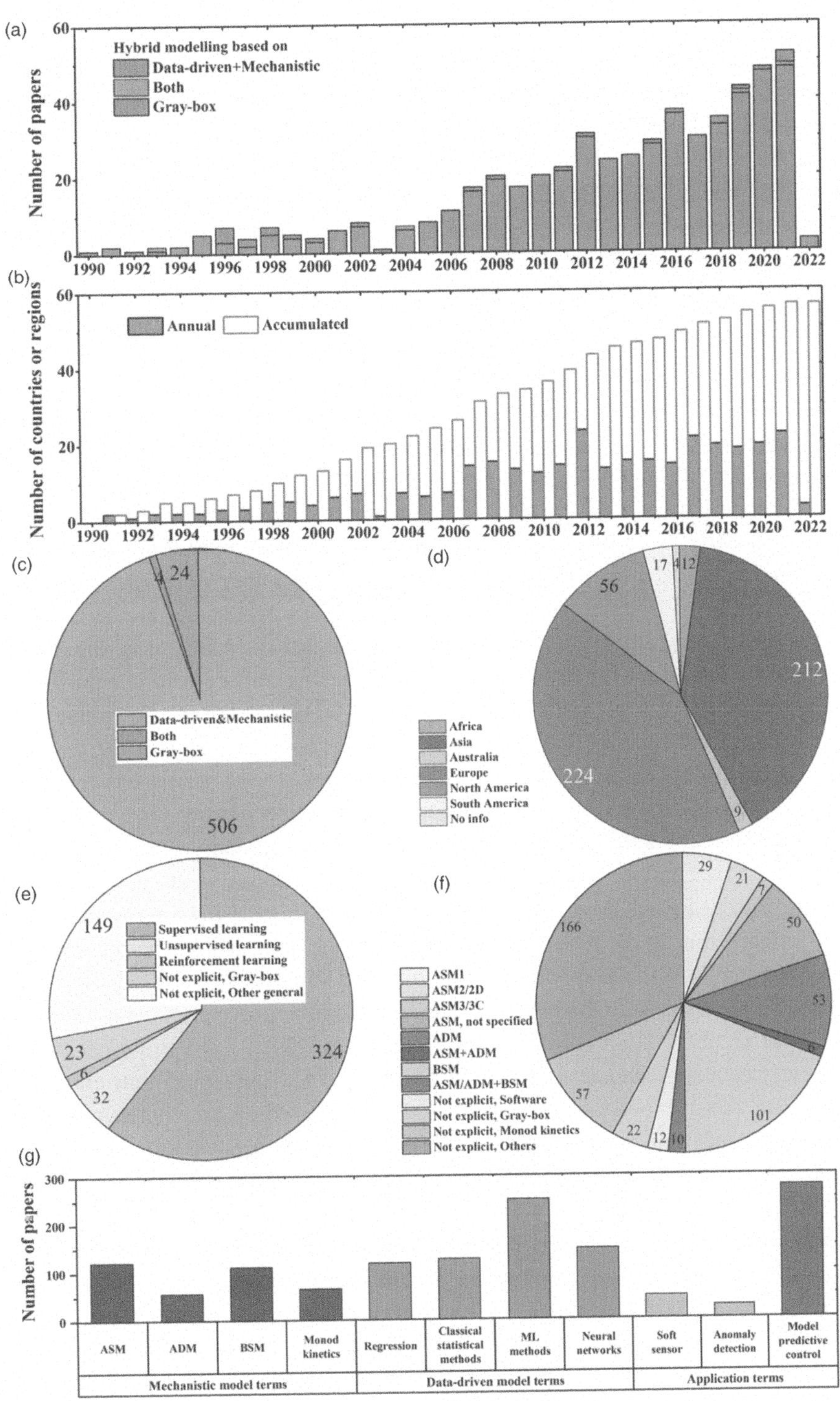

Figure 3 | Text mining analysis of the 534 relevant papers focused on HM research in wastewater management (data updated on January 1, 2022). Temporal variations in (a) the number of papers and (b) the number of countries or regions (annual and accumulated) associated with the papers published from 1990 to 2021. The numbers of HM papers that (c) were identified based on data-driven and mechanistic methods, gray-box, and both; (d) were authored from different areas based on the corresponding authors; (e) used supervised learning (neural networks are included here), unsupervised learning, reinforcement learning, not explicit but used gray-box, and other not explicit; and (f) studied ASM/ADM, BSM, and various not explicit (simulation software, gray-box, Monod kinetics, and others); (g) number of papers associated with 11 selected terminologies in the relevant papers.

3.1. HM in wastewater treatment

ASM mainly dominates WRRF modelling for the primary biological process, along with other unit processes such as sedimentation. ASM1, ASM2d, and ASM3 (Henze *et al.* 2000) have provided a standard framework for modelling based on which several other variants are developed. The use of hybrid models in wastewater treatment is also extensively documented in the literature (e.g. Boger 1992; Côté *et al.* 1995), long before the utilisation of ML methods and the current digitalisation trends related to the so-called, 4th Industrial Revolution (Corominas *et al.* 2018). Due to the increasing number of studies focusing on HM of WRRFs, the primary aspects covered in the existing studies are broadly mentioned below, emphasising the role of HM compared to using only mechanistic or data-driven models.

- **Influent generation:** The availability of dynamic influent data is a major bottleneck for applying ASMs to evaluate the design and operational scenarios for WRRFs. The WRRF influent generation concerns many complex processes, i.e. from at-source to the outlet of the sewer system. Hence, the development of influent generators focusing on the wastewater dynamics at the entrance of WRRFs, which can also be coupled with ASMs, continues to be a topic of interest. Influent generator models that use phenomenological or empirical approaches (De Keyser *et al.* 2010; Gernaey *et al.* 2011), as well as ML methods (Ahnert *et al.* 2016; Zhu *et al.* 2018; Ansari *et al.* 2020) based on historical data, are widely available. In addition, limited hybrid models are also developed that combine conceptual and stochastic models for dynamic influent generation (Talebizadeh *et al.* 2016). Such models aim to improve the model predictability for the key input state variables when compared to solely using a mechanistic or data-driven approach. Another generally employed approach is to introduce randomness or stochasticity to the predicted outputs using phenomenological methods to replicate real data (De Keyser *et al.* 2010; Flores-Alsina *et al.* 2014; Martin & Vanrolleghem 2014). In this case, the aim is to generate realistic influent data that is especially important for dynamic process modelling and control design. Influent generation is not only limited to predicting the incoming flow rate and pollutant loads to WRRF but is also valuable for generating dynamic inputs from other sources like domestic households (Penn *et al.* 2017; Sitzenfrei *et al.* 2017; Wärff *et al.* 2020) and combined sewer overflows (Keupers & Willems 2015).

- **Process modelling:** Hybrid models can be used to fill the gaps in prediction performance and process knowledge arising mainly from the use of mechanistic models (Anderson *et al.* 2000). Several serial approaches are successfully demonstrated, where ANNs and their variants are commonly used (e.g. Côté *et al.* 1995; Cong *et al.* 2020; Cheng *et al.* 2021). Studies using other approaches like Gaussian processes (Hvala & Kocijan 2020) and those using an ensemble of several methods (Lee *et al.* 2005) are also performed. Other studies use parallel HM, where the underlying dynamics and the relevant process are modelled based on the ASM family, and a data-driven model is used to improve the prediction capability (Thompson & Kramer 1994; Lee *et al.* 2005; Villez *et al.* 2020). HM is further applied to improve parameter identification in process modelling using ASM (Mašić *et al.* 2017; Villez *et al.* 2019). Zhu *et al.* (2015) and Borzooei *et al.* (2019) developed a HM approach to improve weather resilience and energy efficiency of a biological nutrient removal activated sludge system at a large-scale WRRF. An unsupervised clustering method was used on an extensive historical climate dataset to identify a series of weather-based influent scenarios, which were further used as an input of the process simulation model. The authors reported that the implementation of weather-based aeration strategies improved the energy efficiency of a WRRF.

- **Process monitoring and control:** HM can support greater plant automation via online, real-time control by taking advantage of the low computational cost of data-driven models. A HM framework can be realised by applying adaptive plant models through the use of digital twins, for example, for the training of surrogate models to be used in model-based control algorithms. Also, hybrid models can present an important added value for other parts of the control loop development. For instance, it is known that data-driven models (such as multilayer perceptrons (MLP)), long short-term memory (LSTM), ARIMA) have excellent performance in time-series prediction (Fernández *et al.* 2009; Zhu *et al.* 2018) and fault/anomaly detection (Wade *et al.* 2005; Haimi *et al.* 2016; Cheng *et al.* 2019). As a result, the forecasting capability of soft sensors can enhance process monitoring when working in parallel with traditional measurements or even reduce the necessity of some direct measurements (Haimi *et al.* 2013). A soft sensor is a model that predicts a variable of interest using the input from physical sensor(s). Soft sensors can predict either real-time/future influent conditions (Wang *et al.* 2019; Zhu & Anderson 2019) or the effluent quality/system performance (Shi & Xu 2018; Niu *et al.* 2020), which can be then used as input information to the feedback controller (utilised with the ASM/BSM models).

Overall, with the increasing model complexity of mechanistic models and the growing acceptance of data-driven approaches, research and application of hybrid modelling approaches to real-world case studies are expected to increase in the future (Sin & Al 2021).

3.2. HM in water treatment

As the wastewater treatment sector is transitioning to a more circular economy, decentralized treatment and water reuse concepts are gaining more attention. It thus becomes more and more challenging to separate wastewater treatment and water treatment into two completely unconnected domains. Hence, to provide a comprehensive overview of the applications and potential of HM in current and future (waste)water treatment processes, we briefly touch upon HM in drinking and process water treatment modelling studies.

The treatment of surface and groundwater resources to maintain a certain level of potable water quality imposed by regulators is based on a wide range of unit operations in water treatment plants (WTPs). Considering the importance of the continuous and efficient operation of WTPs in public health, several phenomenological models have been developed to simulate water treatment processes (Bhattacharjee & Tollner 2016). These models provide mathematical interpretations of various mechanisms, including the flow of water through reactor, equilibrium between water and gas, or solid phases in, for example, flotation processes (Yang *et al.* 2021), decay in the water and/or solid-phase in for example disinfection processes (Crapulli *et al.* 2014), chemical reactions (e.g., coagulation processes), and solution-diffusion processes in for example membrane processes (Gaublomme *et al.* 2020). Mechanistic models include programs developed for individual treatment processes as well as a few simulation packages providing the plant-wide simulation of WTPs such as Stimela (Van der Helm & Rietveld 2002), OTTER (Head *et al.* 2002), Metrex (Mälzer & Nahrstedt 2002), and TAPWAT (Versteegh *et al.* 2001). One of the main limitations for more widespread use of mechanistic models is that some of the input and/or calibration parameters of several unit process models, such as floc porosity, colloid charge, micropollutant concentration, and pathogen abundance, are unknown, case-specific, and hard to measure with sufficient precision (Juntunen *et al.* 2013; Gomes *et al.* 2015).

On the other hand, several studies applied data-driven approaches to monitor, model, and predict source water quality (e.g., Sahoo *et al.* 2005; Stedmon *et al.* 2011; Perelman *et al.* 2012; Mei *et al.* 2014; Debnath *et al.* 2015; Keskin *et al.* 2015; Deng & Wang 2017; Tesoriero *et al.* 2017; Mohammed *et al.* 2018; Delpla *et al.* 2019; Guo *et al.* 2019; Jin *et al.* 2019; Panidhapu *et al.* 2020; Li & Peleato 2021) and to model and control water treatment operations (e.g., Gagnon *et al.* 1997; Maier *et al.* 2004; Chen & Hou 2006; Wu & Lo 2008; Griffiths & Andrews 2011; Heddam *et al.* 2012; Juntunen *et al.* 2013; Gomes *et al.* 2015). Parallel to the increasing interest in using data-driven models, concerns have been raised about the liability and accountability of these models implemented in safety-critical systems such as water treatment, in which the slightest malfunction and failures may damage human health and/or the environment (Aliashrafi *et al.* 2021).

Recent studies have investigated the combination of data-driven models or hybridisation in response to explainability, auditability, and interpretability requirements (Doshi-Velez & Kim 2017; Abba *et al.* 2020) of data-driven approaches and first principles for water treatment processes (Karniadakis *et al.* 2021). An integrated framework was proposed in Wang *et al.* (2016), combining process-based and data-driven models to improve the forecasting of salinity and dissolved oxygen parameters for a real case study. The hybrid platform consisted of two methods based on chaos theory coupled with water quality models. It was reported that hybridization could improve the forecast ability of the model and reduce the computation time. Jia *et al.* (2021) investigated the application of a physics-guided recurrent neural network in the context of modelling the dynamics of the temperature of water resources. It was reported that hybridization of the law of energy conservation with LSTM could improve the forecasting accuracy of the model. By applying the hybrid linear-state-space dynamic Langmuir model, Nordstrand & Dutta (2021) achieved a more robust and flexible control of capacitive-deionisation desalination processes. The authors reported that combining a general modelling framework with physical insights could offer flexibility in calibration data and better predictability over a wide range of operating models.

It should be noted that digital adaptation has been far quicker in the water treatment sector than in wastewater, mainly due to more straightforward functionality and higher reliability of water quantity measuring devices used in WTPs than wastewater quality sensors in WRRFs (Therrien *et al.* 2020). However, to the authors' knowledge, only a few studies have been found implementing a HM approach in WTPs. This can be since there is still a lack of a standardised mechanistic modelling framework similar to ASM models in water treatment processes. In contrast, the water treatment sector has employed a wide variety of mechanistic and data-driven models to address the modelling needs. This might influence the development of hybrid models in these two sectors.

4. CHALLENGES AND DEVELOPMENT NEEDS

By combining the fundamental and generalisable process knowledge built into mechanistic models and the predictive power of data-driven approaches, HM offers an opportunity to overcome inherent weaknesses in both and significantly improve the overall model performance. However, the application of HM as a powerful tool for application in WRRF modelling studies is hampered by several critical challenges and associated development needs, which are discussed in this section.

4.1. Good modelling practice frameworks for HM

The WRRF community has a strongly developed modelling community (evidenced by the organisation of specialist WRRmod seminars and the IWA Specialist Group on Modelling and Integrated Assessment, MIA), which has stimulated the development of robust modelling frameworks and protocols within the domain. Standardised model frameworks, such as ASM, have been used for decades, and different protocols for Good Modelling Practice (GMP), e.g., for calibration and validation (Rieger *et al.* 2012) or uncertainty analysis (Belia *et al.* 2021), have been suggested to encourage a systematic workflow for model development and application. Integrating new modelling paradigms into existing WRRF modelling frameworks raises several questions on extending the existing protocols and practices to continue compliance with GMP guidelines. Azarpour *et al.* (2017) suggest GMP be used for hybrid model identification, but the application of these guidelines should be assessed within the context of existing frameworks for WRRF modelling. Questions to be answered in future research include: Are current protocols for WRRFs transferable to HM, or does the addition of a data-driven component require separate considerations? Should the data-driven and mechanistic components be considered together for each sub-process (e.g., hydraulics, aeration, sludge balance), or should the calibration first focus on the mechanistic models before including any data-driven components?

4.2. How to deal with uncertainty?

Simulation studies are sensitive to different sources of uncertainty, i.e. input data quantity and quality, inadequate model structure, inaccurately estimated model parameters and the inherent uncertainty of the system (Belia *et al.* 2013). Some uncertainty estimation methods are available, such as Monte Carlo and Generalised Likelihood Uncertainty Estimation (GLUE). The sources of uncertainty in HM originate, naturally, from both the mechanistic and data-driven components. However, many previous HM studies lack the quantification of the inherent uncertainty. This makes it difficult to assess whether the model prediction is supported by the training data or due to an uncertain extrapolation. In HM, the accuracy and uncertainty of the results depend on the data quantity and quality used for training/calibration (Refsgaard *et al.* 2007). This means that when process data changes, then the uncertainty of the results will change. The uncertainty quantification is an example of why a consistent collection of meta-data is essential, for example, to consider the wear-and-tear of sensor technologies (Ohmura *et al.* 2019). Dynamically quantifying uncertainty in the context of HM is an essential challenge to assess the trust in hybrid models and, thus, to fully establish and benefit from these powerful tools.

4.3. Model selection – balancing complexity

Determining optimal model complexity for a specific objective is equally relevant for both mechanistic and data-driven models. An overly complex model makes parameter identification impractical or even infeasible for mechanistic models and increases computational cost, rendering the model less user-friendly. Depending on the system and the intended purpose, the modeller must often strike a careful balance between model complexity and usability. An example of this is the wide range of existing models to describe mixing behaviour in a reactor. They range from a simple continuously stirred tank reactor (CSTR) that assumes instant mixing and tanks-in-series and compartmental models that simplify mixing behaviour to advanced computational fluid dynamics (CFD) models, which explicitly model all underlying mixing dynamics. A CFD model would be impractical for most real-time applications, while a CSTR might not yield sufficiently detailed results for an in-depth study.

Similarly, for data-driven models, an appropriate level of complexity needs to be carefully considered. In data-driven terminology, this is known as the bias-variance trade-off. If the model is not complex enough, it cannot capture the relationships, resulting in the underfitting of the model with a high bias. As the complexity of a model is increased, more complex relationships can be considered. For example, a simple linear regression model can only fit a straight line, while a polynomial can fit to nonlinear dynamics. It is not only possible to increase the complexity by switching between techniques; most methods include certain hyperparameters controlling the complexity, e.g., the number of layers and neurons in a neural network; however, as complexity increases, the probability of fitting to noise increases (Fayyad *et al.* 1996). Consequently, the model learns a relationship that it cannot generalise to new data.

The application of HM brings the additional challenge of balancing complexity between the mechanistic and data-driven components. For example, in the serial HM approach, there is a risk that one model compensates for a suboptimal structure of the other model. Since mechanistic and data-driven modelling paradigms are rooted in different engineering practices, only a few experts are trained to a sufficient level in both, which might lead to sub-optimally balanced models, biased to the methods most comfortable with the practitioner. A sub-optimally balanced model is one in which the prediction power is lower than its potential prediction capacity. For example, the data-driven model will try to compensate for the lack of identifiability in the mechanistic model, thus hampering its ability to accurately capture the process dynamics for the subprocess for which it is being developed.

4.4. Data quality and metadata

A fundamental prerequisite for the data-driven part of the HM application is ensuring a high-quality data pipeline to avoid a 'garbage in – garbage out' scenario (Therrien *et al.* 2020; Dobbelaere *et al.* 2021). We put special emphasis on the data quality because, e.g., if there is an operational change from routine, manual sampling at a laboratory to the use of an online sensor, then a wholesale change in protocol for screening, processing and analysing the data is likely necessary. However, Schneider *et al.* (2019) and Thürlimann *et al.* (2019) demonstrate that inaccurate data can be beneficial, for example, in soft-sensor development, where choosing an approach robust to the disturbances will require knowledge of uncertainties and errors in the data. An additional challenge is descriptive data, commonly referred to as meta-data. Especially for large amounts of data, meta-data is invaluable and necessary to ensure the data and metadata are in a consistent, machine-interpretable form, preferably with an automated collection (Aguado *et al.* 2021). Automatic collection forces a clear definition and helps to avoid gaps in the metadata collection.

4.5. Architecture challenge

As discussed in section 2, HM can be constructed according to different architectures and choosing the right one is an important step in the construction of HM frameworks. Choosing which architecture to use is mainly problem-dependent, conditional on the data available and the knowledge missing. Given that theory lags behind the practice in ML, trial and error are a large part of a model's development process. According to von Stosch *et al.* (2014), a serial architecture is preferred in cases where the confidence in the overall mechanistic model structure is relatively high, but specific subprocesses are insufficiently known or understood. One could argue that this is precisely the case for WRRF modelling. The ASM provide a backbone model structure, but modellers often struggle to accurately capture the variability in specific subprocesses such as microbial composition (Ni & Yu 2010), mixing behaviour (Arnaldos *et al.* 2015), or oxygen transfer over a wide range of operating conditions (Amaral *et al.* 2019). A nice example is a recent effort to develop ASM-based model extensions to capture the dynamics of short cut N-removal processes (Sharif Shourjeh *et al.* 2021) or efforts to increase the realism of mixing behaviour deriving compartmental models from CFD models (Le Moullec *et al.* 2011). HM could provide an interesting solution here as they present benefits derived from knowledge incorporated in existing mechanistic wastewater modelling frameworks, such as ASM/ADM, while capturing unobserved dynamics flexibly. An important challenge for such applications of serial hybrid models is that direct and frequent measurements of the relevant variables for the unknown subprocesses (such as microbial composition, dynamic oxygen transfer efficiency, specific growth rates) are often not possible. Thus, the data-driven component of the hybrid model in such a case can only be trained indirectly, or the hybrid model should be calibrated as a whole on available data through the use of, e.g., neural differential equations. Benchmarking the potential of HM for several commonly known ill-defined subprocesses is an interesting and pragmatic avenue to advance the uptake of HM in the water and resource domain.

4.6. The reputation that data-driven methods impede process understanding

The data-driven approach is often associated with a lack of interpretability, such as for SVM (Angelov *et al.* 2021), but it is not always the case. Data-driven modelling can support the understanding and interpretation of data. Examples are dimension reduction methods such as self-organising maps (Dürrenmatt & Gujer 2011) or PCA (Lee & Vanrolleghem 2003) in combination with clustering methods (Villez *et al.* 2008). Furthermore, explainable AI (Samek *et al.* 2017; Adadi & Berrada 2018) or decision trees (Quinlan 1990) are data-driven methods with higher interpretability (Angelov *et al.* 2021). Hybrid models suffer much less from the lack of interpretability. Since the data-driven part is inherently coupled to a model structure with physical meaning, this allows us to evaluate the specific dynamics of the data-driven part as a measure of model structure error in the mechanistic model. Such hybrid models have the potential to provide insights into ill-defined processes. Therefore, applying interpretable ML tools such as Shapley values (Shapley 2016; Lundberg & Lee 2017) to derive mechanistic insights from HM deserves attention in future research.

4.7. Common API and open source software

For hybrid models to be used to their full potential as standard practice for WRRF modelling, it is imperative to have access to a compatible platform or application programming interface (API). However, the data-driven approaches are often hard coded using open-source software, e.g., Python, R, MATLAB, Julia, while mechanistic models, such as ASM and the ADM No. 1, have been packaged in off-the-shelf commercial simulators. Some WRRF simulation software providers (Sumo, WEST, Simba#, GPS-X, dDockDT) embrace open data philosophy and build APIs for open-source scripting languages. Improvement of the compatibility and flexibility of these simulation platforms is still in progress. Furthermore, maintaining the pre-release research code is an issue, especially when linking several different software. As many developers leave academia, a similar effort might be required several times.

5. VISION ON HYBRID MODELS AS AN ENABLING FACTOR IN THE TRANSITION TOWARDS A MORE RESOURCE-EFFICIENT WATER SECTOR

There has been a clear trend of increasing utilisation of data-driven models in the water sector over the last two decades (Eggimann *et al.* 2017). Many data-driven approaches have the potential to break through and become routinely used tools in the industry or have been ranked as an important topic for research and development by experts in the WRRF field (Blumensaat *et al.* 2019). Some data-driven models have replaced, or will soon replace, more conventional knowledge-based models (Hadjimichael *et al.* 2016). Much research exists on the application of ML in the urban water sector. However, mechanistic models will remain necessary as long as high-quality process data is laborious, difficult, and costly to collect and significant uncertainty remains unaccounted for in treatment processes. On the other hand, data-driven approaches are viewed with scepticism due to the lack of process explainability (Newhart *et al.* 2019). However, we see a huge potential for hybrid model development and application due to the combination of full or partial mechanistic explainability, predictive power, and computational efficiency. As such, HM can lead to significant process improvements, especially in the context of multi-objective decision-making, which is already a significant challenge for WRRF operators. Specifically, we see a large potential for the application of HM in the following areas.

5.1. Soft-sensor development for decentralised water treatment and reuse

WRRF have an important societal function to make urban water management more resource-efficient, sustainable, reliable, and safe (Larsen 2011; Sedlak 2014; van Loosdrecht & Brdjanovic 2014). Decentralised WRRF play an important role to reach this goal as they stimulate a more circular water economy through the production of water fit for purpose (Larsen *et al.* 2016). However, a true paradigm shift is only possible with appropriate monitoring, control, and optimization (Eggimann *et al.* 2017). Unfortunately, the amount of undetected failures at decentralised WRRFs is significant (Moelants *et al.* 2008), which results in reduced performance. To monitor such plants, soft sensors (which provide an estimation of variables that are difficult to measure automatically due to sensor availability and cost limitations), using data from low-cost, unmaintained, and likely faulty sensors (80% accuracy in a real-world implementation) can be developed (Schneider *et al.* 2020). There is, however, still significant scope for improvement of these types of soft sensors. We think that HM can improve the acuity and reliability of soft sensors and can be used more generally at unsupervised WRRF, for example. Additionally, a parallel HM approach can be considered to increase the robustness of the monitoring, especially for plants where an undetected failure has a critical process impact. Moreover, a more widespread use, or optimal placement, of sensors (Villez *et al.* 2020) will allow the collection of more valuable datasets that can be utilised to populate the data-driven component of hybrid models with higher resolution and better quality data.

5.2. Harnessing the power of advanced, computationally expensive modelling frameworks

Additionally, HM can facilitate the transition of advanced model frameworks to the WRRF modelling practice. Examples are models describing particle heterogeneity, such as population balance models (Nopens *et al.* 2015). This model framework has the theoretical potential to describe many of the dynamics related to particle heterogeneity in WRRF processes, such as floc and granule formation, struvite crystal growth, or precipitation (Nopens *et al.* 2015; Elduayen 2020) but is dependent on functions that describe sub-processes such as growth, aggregation, and break-up for which current mechanistic knowledge is insufficient. HM can be an enabling factor for the application of these models by combining population balance models (to gain insight into processes driven by variables with important distributed properties, e.g., microbial diversity, floc size and structure, precipitation) with data-driven models to describe the sub-processes (e.g., particle interactions) (Torfs *et al.* 2012).

5.3. Advanced control and optimisation of urban water systems

Recent trends in the digitalisation of the water sector have led to the emergence of digital twins as powerful decision-making tools. A digital twin is a virtual representation of a physical entity (i.e. a model) that has a live, automated data connection to its physical counterpart (Fuller *et al.* 2020). As such, a digital twin allows for process simulations in (near) real-time, opening up the potential for real-time operational decision-making, advanced process control, and online process optimisation at a scale beyond existing automation to date. The use of digital twins brings along an important shift in model objectives. For real-time scenario analysis or control, operational digital twins require a very high predictive power, typically on a short time-scale (hours to days in the future) (Wright & Davidson 2020). Assuming sufficient high-quality data is available, this time frame is exactly the ecosystem in which data-driven models can perform very well. However, despite the consensus on their great potential and substantial research efforts, purely data-driven models are not yet widely used for full-scale WRRF operations. This is due to several factors such as the non-stationary behaviour of these systems, sizeable daily flow fluctuations (Lowe *et al.* 2010), variations in sludge residence time or dilution effects. Hybrid models have the potential to boost the predictive power in real-time applications by leveraging the power of data-driven techniques without losing the process knowledge incorporated in the mechanistic model. An example is the use of reinforcement learning or active learning (Hernández-del-Olmo *et al.* 2018; Filipe *et al.* 2019; Alves Goulart & Dutra Pereira 2020), which will bring benefits to better deployment of real-time control. This can accelerate the transition of WRRF models to digital twin applications and, as such, stimulate real-time and online optimisation of WRRF concerning energy consumption, effluent quality (tailored for discharge or specific reuse purposes), or greenhouse gas emissions.

5.4. Beyond the combination of two paradigms

Most current applications of HM are either based on the development of surrogate models or the traditional serial or parallel architecture. In the near future, such approaches have a great potential to improve the management of WRRF significantly. However, in these approaches, the contributions of the data-driven and mechanistic components are still relatively separated. Taking a step forward, we propose truly integrated approaches rather than combining two distinct models, whether serially or in parallel. This means an iterative approach in which one model profits and learns from the other internally and intrinsically. For example, it can be used to explore a wide range of potential mechanistic model pathways. An unknown process could be evaluated first using a data-driven methodology. With a human-in-the-loop approach, information can be checked for consistency and acuity prior to accepting the knowledge gained from the model. This information would then flow into the design for the next experiment to gain more detailed or structured mechanistic knowledge. Hence, in addition to improving the prediction power, greater process understanding can be obtained at a relatively lower cost than by mechanistic modelling alone.

ACKNOWLEDGEMENTS

We thank Kris Villez for critically reviewing the manuscript and all the HM workshop participants at the WRRmod 2021 conference in Arosa, Switzerland, for input and discussion.

FUNDING SOURCES

MYS received funding from the Japanese Society for Promotion of Science (JSPS) Grant P20763. WQ received funding from the Research Foundation – Flanders (FWO) through Grant 3S79219.

AUTHOR CONTRIBUTIONS

All authors contributed equally to the development and finalization of this manuscript.

DATA AVAILABILITY STATEMENT

All relevant data are included in the paper or its Supplementary Information.

REFERENCES

Abba, S. I., Pham, Q. B., Saini, G., Linh, N. T. T., Ahmed, A. N., Mohajane, M., Khaledian, M., Abdulkadir, R. A. & Bach, Q.-V. 2020 Implementation of data intelligence models coupled with ensemble machine learning for prediction of water quality index. *Environ. Sci. Pollut. Res.* **27**, 41524–41539.

Abrahart, R. J., See, L. M. & Solomatine, D. P. 2008 *Practical Hydroinformatics: Computational Intelligence and Technological Developments in Water Applications.* Springer Science & Business Media, Berlin, Heidelberg, Germany.

Adadi, A. & Berrada, M. 2018 Peeking inside the black-box: a survey on explainable artificial intelligence (XAI). *IEEE Access* **6**, 52138–52160. https://doi.org/10.1109/ACCESS.2018.2870052.

Aguado, D., Blumensaat, F., Baeza, J. A., Villez, K., Ruano, M. V., Samuelsson, O. & Plana, Q. 2021 *Digital Water, The Value of Meta-Data for Water Resource Recovery Facilities.* IWA.

Ahnert, M., Marx, C., Krebs, P. & Kuehn, V. 2016 A black-box model for generation of site-specific WWTP influent quality data based on plant routine data. *Water Science and Technology* **74** (12), 2978–86. https://doi.org/10.2166/wst.2016.463.

Aliashrafi, A., Zhang, Y., Groenewegen, H. & Peleato, N. M. 2021 A review of data-driven modelling in drinking water treatment. *Rev. Environ. Sci. Biotechnol.* **20**, 985–1009.

Alves Goulart, D. & Dutra Pereira, R. 2020 Autonomous pH control by reinforcement learning for electroplating industry wastewater. *Comput. Chem. Eng.* **140**, 106909. https://doi.org/10.1016/j.compchemeng.2020.106909.

Amaral, A., Gillot, S., Garrido-Baserba, M., Filali, A., Karpinska, A. M., Plósz, B. G., De Groot, C., Bellandi, G., Nopens, I., Takács, I., Lizarralde, I., Jimenez, J. A., Fiat, J., Rieger, L., Arnell, M., Andersen, M., Jeppsson, U., Rehman, U., Fayolle, Y., Amerlinck, Y. & Rosso, D. 2019 Modelling gas–liquid mass transfer in wastewater treatment: when current knowledge needs to encounter engineering practice and vice versa. *Water Sci. Technol.* **80**, 607–619. https://doi.org/10.2166/wst.2019.253.

Angelov, P. P., Soares, E. A., Jiang, R., Arnold, N. I. & Atkinson, P. M. 2021 Explainable artificial intelligence: an analytical review. *WIREs Data Min. Knowl. Discovery* **11**, e1424. https://doi.org/10.1002/widm.1424.

Ansari, M., Othman, F. & El-Shafie, A. 2020 Optimized fuzzy inference system to enhance prediction accuracy for influent characteristics of a sewage treatment plant. *Science of The Total Environment* **722**, 137878. https://doi.org/10.1016/j.scitotenv.2020.137878.

Anderson, J. S., McAvoy, T. J. & Hao, O. J. 2000 Use of hybrid models in wastewater systems. *Industrial & Engineering Chemistry Research* **39** (6), 1694–1704. https://doi.org/10.1021/ie990557r.

Apache Software Foundation n.d. *Hadoop (Version 3.3.1).* Available from: https://hadoop.apache.org. (accessed 1 July 2022).

Arnaldos, M., Amerlinck, Y., Rehman, U., Maere, T., Van Hoey, S., Naessens, W. & Nopens, I. 2015 From the affinity constant to the half-saturation index: understanding conventional modeling concepts in novel wastewater treatment processes. *Water Res.* **70**, 458–470. https://doi.org/10.1016/j.watres.2014.11.046.

Azarpour, A., Borhani, T. N. G., Wan Alwi, S. R., Manan, Z. A. & Abdul Mutalib, M. I. 2017 A generic hybrid model development for process analysis of industrial fixed-bed catalytic reactors. *Chem. Eng. Res. Des.* **117**, 149–167. https://doi.org/10.1016/j.cherd.2016.10.024.

Baeten, J. E., Batstone, D. J., Schraa, O. J., van Loosdrecht, M. C. M. & Volcke, E. I. P. 2019 Modelling anaerobic, aerobic and partial nitritation-anammox granular sludge reactors – a review. *Water Res.* **149**, 322–341. https://doi.org/10.1016/j.watres.2018.11.026.

Belia, E., Johnson, B., Benedetti, L., Bott, C. B., Martin, C., Murthy, S., Neumann, M. B., Rieger, L., Weijers, S. & Vanrolleghem, P. A. 2013 *Uncertainty Evaluations in Model Based WRRF Design for High Level Nutrient Removal.* WERF NUTR1R06q, 54.

Belia, E., Neumann, M. B., Benedetti, L., Johnson, B., Murthy, S., Weijers, S. & Vanrolleghem, P. A. 2021 *Uncertainty in Wastewater Treatment Design and Operation.* IWA Publishing.

Belia, Evangelia, Lorenzo Benedetti, Bruce Johnson, Sudhir Murthy, Marc Neumann, Peter Vanrolleghem, and Stefan Weijers, eds. Uncertainty in Wastewater Treatment Design and Operation: Addressing Current Practices and Future Directions. Scientific and Technical Report 21. London, UK: IWA Publishing, 2021. https://doi.org/10.2166/9781780401034.

Bhattacharjee, N. V. & Tollner, E. W. 2016 Improving management of windrow composting systems by modeling runoff water quality dynamics using recurrent neural network. *Ecol. Modell.* **339**, 68–76.

Blumensaat, F., Leitão, J. P., Ort, C., Rieckermann, J., Scheidegger, A., Vanrolleghem, P. A. & Villez, K., 2019 How urban storm- and wastewater management prepares for emerging opportunities and threats: digital transformation, ubiquitous sensing, new data sources, and beyond – a horizon scan. *Environ. Sci. Technol.* acs.est.8b06481. https://doi.org/10.1021/acs.est.8b06481

Boger, Z. 1992 Application of neural networks to water and wastewater treatment plant operation. *ISA Trans.* **31**, 25–33. https://doi.org/10.1016/0019-0578(92)90007-6.

Borzooei, S., Miranda, G. H., Teegavarapu, R., Scibilia, G., Meucci, L. & Zanetti, M. C. 2019 Assessment of weather-based influent scenarios for a WWTP: application of a pattern recognition technique. *J. Environ. Manage.* **242**, 450–456.

Borzooei, S., Miranda, G. H., Abolfathi, S., Scibilia, G., Meucci, L. & Zanetti, M. C. 2020 Application of unsupervised learning and process simulation for energy optimization of a WWTP under various weather conditions. *Water Sci. Technol.* **81**, 1541–1551.

Butler, D., Ward, S., Sweetapple, C., Astaraie-Imani, M., Diao, K., Farmani, R. & Fu, G. 2017 Reliable, resilient and sustainable water management: the Safe & SuRe approach. *Global Challenges* **1** (1), 63–77. https://doi.org/10.1002/gch2.1010.

Cecconi, F., Reifsnyder, S., Sobhani, R., Cisquella-Serra, A., Madou, M. & Rosso, D. 2020 Functional behaviour and microscopic analysis of ammonium sensors subject to fouling in activated sludge processes. *Environ. Sci. Water Res. Technol.* **6**, 2723–2733. https://doi.org/10.1039/D0EW00359 J.

Chen, C.-L. & Hou, P.-L. 2006 Fuzzy model identification and control system design for coagulation chemical dosing of potable water. *Water Sci. Technol. Water Supply* **6**, 97–104.

Chen, R. T., Rubanova, Y., Bettencourt, J. & Duvenaud, D. 2018 *Neural Ordinary Differential Equations*. ArXiv Prepr. ArXiv180607366.

Cheng, G., Ramirez-Amaro, K., Beetz, M. & Kuniyoshi, Y. 2019 Purposive learning: robot reasoning about the meanings of human activities. *Sci. Rob.* **4**. https://doi.org/10.1126/scirobotics.aav1530.

Cheng, X., Guo, Z., Shen, Y., Yu, K. & Gao, X. 2021 Knowledge and data-driven hybrid system for modeling fuzzy wastewater treatment process. *Neural Comput. Appl.* https://doi.org/10.1007/s00521-021-06499-1.

Cong, Q., Xi, X., Su, C., Deng, S. & Zhao, Y. 2020 Hybrid integrated model of water quality in wastewater treatment process via RBF neural network. In: *Robotics and Rehabilitation Intelligence, Communications in Computer and Information Science* (Qian, J., Liu, H., Cao, J. & Zhou, D., eds). Springer, Singapore, pp. 333–341. https://doi.org/10.1007/978-981-33-4932-2_24.

Conti, S. & Hagan, A. O. 2010 Bayesian emulation of complex multi-output and dynamic computer models. **140**, 640–651. https://doi.org/10.1016/j.jspi.2009.08.006.

Corominas, L., Garrido-Baserba, M., Villez, K., Olsson, G., Cortés, U. & Poch, M. 2018 Transforming data into knowledge for improved wastewater treatment operation: a critical review of techniques. *Environ. Modell. Software* **106**, 89–103. https://doi.org/10.1016/j.envsoft.2017.11.023.

Côté, M., Grandjean, B. P. A., Lessard, P. & Thibault, J. 1995 Dynamic modelling of the activated sludge process: improving prediction using neural networks. *Water Res.* **29**, 995–1004. https://doi.org/10.1016/0043-1354(95)93250-W.

Crapulli, F., Santoro, D., Sasges, M. R. & Ray, A. K. 2014 Mechanistic modeling of vacuum UV advanced oxidation process in an annular photoreactor. *Water Res.* **64**, 209–225. https://doi.org/10.1016/j.watres.2014.06.048.

Daims, H. 2014 The family nitrospiraceae. In: *The Prokaryotes* (Rosenberg, E., DeLong, E. F., Lory, S., Stackebrandt, E. & Thompson, F., eds). Springer Berlin Heidelberg, Berlin, Heidelberg, pp. 733–749. https://doi.org/10.1007/978-3-642-38954-2_126.

DeAngelis, D. L. & Yurek, S. 2015 Equation-free modeling unravels the behavior of complex ecological systems. *Proc. Natl. Acad. Sci.* **112**, 3856–3857.

Debnath, A., Majumder, M. & Pal, M. 2015 A cognitive approach in selection of source for water treatment plant based on climatic impact. *Water Resour. Manage.* **29**, 1907–1919.

De Keyser, W., Gevaert, V., Verdonck, F., De Baets, B. & Benedetti, L. 2010 An emission time series generator for pollutant release modelling in urban areas. *Environ. Model. Software* **25**, 554–561. https://doi.org/10.1016/j.envsoft.2009.09.009.

Delpla, I., Florea, M. & Rodriguez, M. J. 2019 Drinking water source monitoring using early warning systems based on data mining techniques. *Water Resour. Manage.* **33**, 129–140.

Deng, W. & Wang, G. 2017 A novel water quality data analysis framework based on time-series data mining. *J. Environ. Manage.* **196**, 365–375.

Dobbelaere, M. R., Plehiers, P. P., Van de Vijver, R., Stevens, C. V. & Van Geem, K. M. 2021 Machine learning in chemical engineering: strengths, weaknesses, opportunities, and threats. *Engineering* **7**, 1201–1211. https://doi.org/10.1016/j.eng.2021.03.019.

Dors, M., Simutis, R. & Lübbert, A. 1995 *Hybrid Process Modeling for Advanced Process State Estimation, Prediction, and Control Exemplified in a Production-Scale Mammalian Cell Culture*. ACS Publications.

Doshi-Velez, F. & Kim, B. 2017 Towards a rigorous science of interpretable machine learning. ArXiv:1702.08608 [Cs, Stat]. https://doi.org/10.48550/arXiv.1702.08608.

Dürrenmatt, D. J. & Gujer, W. 2011 Data-driven modeling approaches to support wastewater treatment plant operation. *Environ. Model. Software* **30**, 47–56. https://doi.org/10.1016/j.envsoft.2011.11.007.

Eggimann, S., Mutzner, L., Wani, O., Schneider, M. Y., Spuhler, D., Moy de Vitry, M., Beutler, P. & Maurer, M. 2017 The potential of knowing more: a review of data-driven urban water management. *Environ. Sci. Technol.* **51**, 2538–2553. https://doi.org/10.1021/acs.est.6b04267.

Elduayen, E. B. 2020 *New Mass-Based Population Balance Model Including Shear Rate Effects: Application to Struvite Recovery*. University of Navarra, Donostia.

Fayyad, U., Piatetsky-Shapiro, G. & Smyth, P. 1996 From data mining to knowledge discovery in databases. *AI Mag.* **17**, 37. https://doi.org/10.1609/aimag.v17i3.1230.

Fernández, C., Vega, J. A., Fonturbel, T. & Jiménez, E. 2009 Streamflow drought time series forecasting: a case study in a small watershed in North West Spain. *Stochastic Environ. Res. Risk Assess.* **23**, 1063–1070.

Filipe, J., Bessa, R. J., Reis, M., Alves, R. & Póvoa, P. 2019 Data-driven predictive energy optimization in a wastewater pumping station. *Appl. Energy* **252**, 113423. https://doi.org/10.1016/j.apenergy.2019.113423.

Flores-Alsina, X., Saagi, R., Lindblom, E., Thirsing, C., Thornberg, D., Gernaey, K. V. & Jeppsson, U. 2014 Calibration and validation of a phenomenological influent pollutant disturbance scenario generator using full-scale data. *Water Res.* **51**, 172–185. https://doi.org/10.1016/j.watres.2013.10.022.

Fuller, A., Fan, Z., Day, C. & Barlow, C. 2020 Digital twin: enabling technologies, challenges and open research. *IEEE Access* **8**, 108952–108971. https://doi.org/10.1109/ACCESS.2020.2998358.

Gagnon, C., Grandjean, B. P. & Thibault, J. 1997 Modelling of coagulant dosage in a water treatment plant. *Artif. Intell. Eng.* **11**, 401–404.

Galvanauskas, V., Simutis, R. & Lübbert, A. 2004 Hybrid process models for process optimisation, monitoring and control. *Bioprocess Biosyst. Eng.* **26**, 393–400.

Gaublomme, D., Strubbe, L., Vanoppen, M., Torfs, E., Mortier, S., Cornelissen, E., De Gusseme, B., Verliefde, A. & Nopens, I. 2020 A generic reverse osmosis model for full-scale operation. *Desalination* **490**, 114509. https://doi.org/10.1016/j.desal.2020.114509.

Gernaey, K. V., Flores-Alsina, X., Rosen, C., Benedetti, L. & Jeppsson, U. 2011 Dynamic influent pollutant disturbance scenario generation using a phenomenological modelling approach. *Environ. Modell. Software* **26**, 1255–1267. https://doi.org/10.1016/j.envsoft.2011. 06.001.

Gernaey, K. V., van Loosdrecht, M. C. M., Henze, M., Lind, M. & Jørgensen, S. B. 2004 Activated sludge wastewater treatment plant modelling and simulation: state of the art. *Environmental Sciences and Artificial Intelligence* **19** (9), 763–83. https://doi.org/10.1016/j. envsoft.2003.03.005.

Ghosh, D., Hermonat, E., Mhaskar, P., Snowling, S. & Goel, R. 2019 Hybrid modeling approach integrating first-principles models with subspace identification. *Ind. Eng. Chem. Res.* **58**, 13533–13543.

Gomes, L. S., Souza, F. A. A., Pontes, R. S. T., Fernandes Neto, T. R. & Araújo, R. A. M. 2015 Coagulant dosage determination in a water treatment plant using dynamic neural network models. *Int. J. Comput. Intell. Appl.* **14**, 1550013.

Griffiths, K. A. & Andrews, R. C. 2011 The application of artificial neural networks for the optimization of coagulant dosage. *Water Sci. Technol. Water Supply* **11**, 605–611.

Guo, D., Lintern, A., Webb, J. A., Ryu, D., Liu, S., Bende-Michl, U., Leahy, P., Wilson, P. & Western, A. W. 2019 Key factors affecting temporal variability in stream water quality. *Water Resour. Res.* **55**, 112–129.

Hadjimichael, A., Comas, J. & Corominas, L. 2016 Do machine learning methods used in data mining enhance the potential of decision support systems? A review for the urban water sector. *AI Commun.* **29**, 747–756. https://doi.org/10.3233/AIC-160714.

Haimi, H., Mulas, M., Corona, F. & Vahala, R. 2013 Data-derived soft-sensors for biological wastewater treatment plants: an overview. *Environ. Modell. Software* **47**, 88–107. https://doi.org/10.1016/j.envsoft.2013.05.009.

Haimi, H., Mulas, M., Corona, F., Marsili-Libelli, S., Lindell, P., Heinonen, M. & Vahala, R. 2016 Adaptive data-derived anomaly detection in the activated sludge process of a large-scale wastewater treatment plant. *Eng. Appl. Artif. Intell.* **52**, 65–80.

Hannaford, N. E., Heaps, S. E., Nye, T. M. W., Curtis, T. P., Allen, B., Golightly, A. & Wilkinson, D. J. 2021 *A Sparse Bayesian Hierarchical Vector Autoregressive Model for Microbial Dynamics in a Wastewater Treatment Plant.* ArXiv210700502 Q-Bio Stat.

Head, R., Shepherd, D., Butt, G. & Buck, G. 2002 OTTER mathematical process simulation of potable water treatment. *Water Sci. Technol. Water Supply* **2**, 95–101.

Heddam, S., Bermad, A. & Dechemi, N. 2012 ANFIS-based modelling for coagulant dosage in drinking water treatment plant: a case study. *Environ. Monit. Assess.* **184**, 1953–1971.

Henze, M., Gujer, W., Mino, T. & van Loosdrecht, M. C. M. 2000 *Activated Sludge Models ASM1, ASM2, ASM2d and ASM3.* IWA publishing, Londen, UK.

Hernández-del-Olmo, F., Llanes, F. H. & Gaudioso, E. 2012 An emergent approach for the control of wastewater treatment plants by means of reinforcement learning techniques. *Expert Syst. Appl.* **39**, 2355–2360. https://doi.org/10.1016/j.eswa.2011.08.062.

Hernández-del-Olmo, F., Gaudioso, E., Dormido, R. & Duro, N. 2018 Tackling the start-up of a reinforcement learning agent for the control of wastewater treatment plants. *Knowledge Based Syst.* **144**, 9–15. https://doi.org/10.1016/j.knosys.2017.12.019.

Hu, G., Mao, Z., He, D. & Yang, F. 2011 Hybrid modeling for the prediction of leaching rate in leaching process based on negative correlation learning bagging ensemble algorithm. *Comput. Chem. Eng.* **35**, 2611–2617. https://doi.org/10.1016/j.compchemeng.2011.02.012.

Hvala, N. & Kocijan, J. 2020 Design of a hybrid mechanistic/Gaussian process model to predict full-scale wastewater treatment plant effluent. *Comput. Chem. Eng.* **140**, 106934. https://doi.org/10.1016/j.compchemeng.2020.106934.

Jia, X., Willard, J., Karpatne, A., Read, J. S., Zwart, J. A., Steinbach, M. & Kumar, V. 2021 Physics-guided machine learning for scientific discovery: an application in simulating lake temperature profiles. *ACMIMS Trans. Data Sci.* **2**, 1–26.

Jin, R., Chen, W. & Simpson, T. W. 2001 Comparative studies of metamodelling techniques under multiple modelling criteria. *Struct. Multidiscip. Optim.* **23**, 1–13.

Jin, T., Cai, S., Jiang, D. & Liu, J. 2019 A data-driven model for real-time water quality prediction and early warning by an integration method. *Environ. Sci. Pollut. Res.* **26**, 30374–30385.

Johansen, T. A. & Foss, B. A. 1992 Representing and learning unmodeled dynamics with neural network memories. In: *1992 American Control Conference.* IEEE, pp. 3037–3043, Chicago, IL, USA.

Johansen, Tor A. & Bjarne, A. 1992 Foss. "Representing and Learning Unmodeled Dynamics with Neural Network Memories.". In 1992 American Control Conference, 3037–43. Chicago, IL, USA, 1992. https://doi.org/10.23919/ACC.1992.4792705.

Juntunen, P., Liukkonen, M., Lehtola, M. & Hiltunen, Y. 2013 Cluster analysis by self-organizing maps: an application to the modelling of water quality in a treatment process. *Appl. Soft Comput.* **13**, 3191–3196.

Karniadakis, G. E., Kevrekidis, I. G., Lu, L., Perdikaris, P., Wang, S. & Yang, L. 2021 Physics-informed machine learning. *Nat. Rev. Phys.* **3**, 422–440.

Karpatne, A., Atluri, G., Faghmous, J. H., Steinbach, M., Banerjee, A., Ganguly, A., Shekhar, S., Samatova, N. & Kumar, V. 2017a Theory-guided data science: a new paradigm for scientific discovery from data. *IEEE Trans. Knowl. Data Eng.* **29**, 2318–2331.

Karpatne, A., Watkins, W., Read, J. & Kumar, V. 2017b *Physics-guided Neural Networks (pgnn): An Application in Lake Temperature Modeling*. ArXiv Prepr. ArXiv171011431.

Keskin, T. E., Düğenci, M. & Kaçaroğlu, F. 2015 Prediction of water pollution sources using artificial neural networks in the study areas of Sivas, Karabük and Bartın (Turkey). *Environ. Earth Sci.* **73**, 5333–5347.

Keupers, I. & Willems, P. 2015 CSO water quality generator based on calibration to WWTP influent data. In: *Proceedings of the 10th International Conference on Urban Drainage Modelling*. Québec, pp. 97–104.

Kramer, M. A., Thompson, M. L. & Bhagat, P. M. 1992 Embedding theoretical models in neural networks. In: *1992 American Control Conference*. IEEE, pp. 475–479, Chicago, IL, USA.

Kramer, M. A., Thompson, M. L. & Bhagat, P. M. 1992 "Embedding Theoretical Models in Neural Networks." In: *1992 American Control Conference*, 475–79. Chicago, IL, USA, 1992. https://doi.org/10.23919/ACC.1992.4792111.

Larsen, T. A. 2011 Redesigning wastewater infrastructure to improve resource efficiency. *Water Sci. Technol.* **63**, 2535–2541. https://doi.org/10.2166/wst.2011.502.

Larsen, T. A., Hoffmann, S., Lüthi, C., Truffer, B. & Maurer, M. 2016 Emerging solutions to the water challenges of an urbanizing world. *Science* **352**, 928–933. https://doi.org/10.1126/science.aad8641.

Lazăr, D. C., Avram, M. F., Faur, A. C., Goldiş, A., Romoşan, I., Tăban, S. & Cornianu, M. 2020 The impact of artificial intelligence in the endoscopic assessment of premalignant and malignant esophageal lesions: present and future. *Medicina (Mex.)* **56**, 364.

Lee, D. S. & Vanrolleghem, P. A. 2003 Monitoring of a sequencing batch reactor using adaptive multiblock principal component analysis. *Biotechnol. Bioeng.* **82**, 489–497. https://doi.org/10.1002/bit.10589.

Lee, D. S., Vanrolleghem, P. A. & Park, J. M. 2005 Parallel hybrid modeling methods for a full-scale cokes wastewater treatment plant. *J. Biotechnol.* **115**, 317–328. https://doi.org/10.1016/j.jbiotec.2004.09.001.

Le Moullec, Y., Potier, O., Gentric, C. & Leclerc, J. P. 2011 Activated sludge pilot plant: comparison between experimental and predicted concentration profiles using three different modelling approaches. *Water Res.* **45**, 3085–3097. https://doi.org/10.1016/j.watres.2011.03.019.

Li, Z. & Peleato, N. M. 2021 Comparison of dimensionality reduction techniques for cross-source transfer of fluorescence contaminant detection models. *Chemosphere* **276**, 130064.

Li, B., Taniguchi, D., Gedara, J. P., Gogulancea, V., Gonzalez-Cabaleiro, R., Chen, J., McGough, A. S., Ofiteru, I. D., Curtis, T. P. & Zuliani, P. 2019 NUFEB: A massively parallel simulator for individual-based modelling of microbial communities. *PLOS Comput. Biol.* **15**, e1007125. https://doi.org/10.1371/journal.pcbi.1007125.

Lowe, K. S., Tucholke, M. B., Tomaras, J. M. B., Conn, K., Hoppe, C., Drewes, J. E., McCray, J. E. & Munakata-Marr, J. 2010 *Influent Constituent Characteristics of the Modern Waste Stream from Single Sources*. IWA Publishing. https://doi.org/10.2166/9781780403519.

Lund, N. S. V., Falk, A. K. V., Borup, M., Madsen, H. & Steen Mikkelsen, P. 2018 Model predictive control of urban drainage systems: a review and perspective towards smart real-time water management. *Crit. Rev. Environ. Sci. Technol.* **48**, 279–339. https://doi.org/10.1080/10643389.2018.1455484.

Lundberg, S. M. & Lee, S.-I. 2017 A unified approach to interpreting model predictions. In *Proceedings of the 31st International Conference on Neural Information Processing Systems*. pp. 4768–4777.

Luo, L. & Bao, S. 2018 Knowledge-data-integrated sparse modeling for batch process monitoring. *Chem. Eng. Sci.* **189**, 221–232.

Mahmoodian, M., Carbajal, J. P., Bellos, V., Leopold, U., Schutz, G. & Clemens, F. 2018 A hybrid surrogate modelling strategy for simplification of detailed urban drainage simulators. *Water Resour. Manage.* **32**, 5241–5256. https://doi.org/10.1007/s11269-018-2157-4.

Maier, H. R., Morgan, N. & Chow, C. W. 2004 Use of artificial neural networks for predicting optimal alum doses and treated water quality parameters. *Environ. Modell. Software* **19**, 485–494.

Mälzer, H. J. & Nahrstedt, A. 2002 *Modellierung mehrstufiger Trinkwasseraufbereitungsanlagen mittels eines expertensystem-basierten Simulationsmodells (Metrex) am Beispiel von oberflächenwasser*.

Mannina, G., Di Bella, G. & Viviani, G. 2011 An integrated model for biological and physical process simulation in membrane bioreactors (MBRs). *J. Membr. Sci.* **376**, 56–69. https://doi.org/10.1016/j.memsci.2011.04.003.

Manning, C., Raghavan, P. & Schütze, H. 2008 The term vocabulary and postings lists. In: Mogotsi, I. C. Christopher D. Manning, Prabhakar Raghavan, and Hinrich Schütze (eds), *Introduction to Information Retrieval*. Cambridge University Press, Cambridge.

Martin, C. & Vanrolleghem, P. A. 2014 Analysing, completing, and generating influent data for WWTP modelling: a critical review. *Environ. Modell. Software* **60**, 188–201. https://doi.org/10.1016/j.envsoft.2014.05.008.

Mašić, A., Srinivasan, S., Billeter, J., Bonvin, D. & Villez, K. 2017 Shape constrained splines as transparent black-box models for bioprocess modeling. *Comput. Chem. Eng.* **99**, 96–105. https://doi.org/10.1016/j.compchemeng.2016.12.017.

Mei, K., Liao, L., Zhu, Y., Lu, P., Wang, Z., Dahlgren, R. A. & Zhang, M. 2014 Evaluation of spatial-temporal variations and trends in surface water quality across a rural-suburban-urban interface. *Environ. Sci. Pollut. Res.* **21**, 8036–8051.

Mitchell, T. 1997 *Machine Learning*. New York, NY, USA: McGraw-Hill.

Moelants, N., Janssen, G., Smets, I. & Van Impe, J. 2008 Field performance assessment of onsite individual wastewater treatment systems. *Water Sci. Technol.* **58**, 1–6. https://doi.org/10.2166/wst.2008.325.

Mohammed, H., Hameed, I. A. & Seidu, R. 2018 Comparative predictive modelling of the occurrence of faecal indicator bacteria in a drinking water source in Norway. *Sci. Total Environ.* **628**, 1178–1190.

Mogotsi, I. C. "Christopher D. Manning, Prabhakar Raghavan, and Hinrich Schütze: Introduction to Information Retrieval." Information Retrieval 13, no. 2 (April 1, 2010): 192–95. https://doi.org/10.1007/s10791-009-9115-y.

Newhart, K. B., Holloway, R. W., Hering, A. S. & Cath, T. Y. 2019 Data-driven performance analyses of wastewater treatment plants: a review. *Water Res.* **157**, 498–513. https://doi.org/10.1016/j.watres.2019.03.030.

Nordstrand, J. & Dutta, J. 2021 Flexible modeling and control of capacitive-deionization processes through a linear-state-space dynamic Langmuir model. *Npj Clean Water* **4** (1), 1–7. https://doi.org/10.1038/s41545-020-00094-y.

Ni, B.-J. & Yu, H.-Q. 2010 Mathematical modeling of aerobic granular sludge: a review. *Biotechnol. Adv.* **28**, 895–909. https://doi.org/10.1016/j.biotechadv.2010.08.004.

Niu, G., Yi, X., Chen, C., Li, X., Han, D., Yan, B., Huang, M. & Ying, G. 2020 A novel effluent quality predicting model based on genetic-deep belief network algorithm for cleaner production in a full-scale paper-making wastewater treatment. *J. Cleaner Prod.* **265**, 121787. https://doi.org/10.1016/j.jclepro.2020.121787.

Nopens, I., Torfs, E., Ducoste, J., Vanrolleghem, P. A. & Gernaey, K. V. 2015 Population balance models: a useful complementary modelling framework for future WWTP modelling. *Water Sci. Technol.* **71**, 159–167. https://doi.org/10.2166/wst.2014.500.

Ohmura, K., Thürlimann, C. M., Kipf, M., Carbajal, J. P. & Villez, K. 2019 Characterizing long-term wear and tear of ion-selective pH sensors. *Water Sci. Technol.* **80**, 541–550. https://doi.org/10.2166/wst.2019.301.

Olsson, G. & Newell, B. 1999 *Wastewater Treatment Systems*. London, UK: IWA Publishing.

Oyebamiji, O. K., Wilkinson, D. J., Li, B., Jayathilake, P. G., Zuliani, P. & Curtis, T. P. 2019 Bayesian emulation and calibration of an individual-based model of microbial communities. *J. Comput. Sci.* **30**, 194–208. https://doi.org/10.1016/j.jocs.2018.12.007.

Pang, J., Yang, S., He, L., Chen, Y. & Ren, N. 2019 Intelligent control/operational strategies in WWTPs through an integrated Q-learning algorithm with ASM2d-Guided reward. *Water* **11**, 927. https://doi.org/10.3390/w11050927.

Panidhapu, A., Li, Z., Aliashrafi, A. & Peleato, N. M. 2020 Integration of weather conditions for predicting microbial water quality using Bayesian Belief Networks. *Water Res.* **170**, 115349.

Penn, R., Schütze, M., Gorfine, M. & Friedler, E. 2017 Simulation method for stochastic generation of domestic wastewater discharges and the effect of greywater reuse on gross solid transport. *Urban Water J.* **14**, 846–852. https://doi.org/10.1080/1573062X.2017.1279188.

Perelman, L., Arad, J., Housh, M. & Ostfeld, A. 2012 Event detection in water distribution systems from multivariate water quality time series. *Environ. Sci. Technol.* **46**, 8212–8219.

Peres, J., Oliveira, R. & De Azevedo, S. F. 2001 Knowledge based modular networks for process modelling and control. *Comput. Chem. Eng.* **25**, 783–791.

Psichogios, D. C. & Ungar, L. H. 1992 A hybrid neural network-first principles approach to process modeling. *AIChE J.* **38**, 1499–1511.

Quaghebeur, W., Nopens, I. & De Baets, B. 2021 Incorporating unmodeled dynamics into first-principles models through machine learning. *IEEE Access* **9**, 22014–22022. https://doi.org/10.1109/ACCESS.2021.3055353.

Quinlan, J. R. 1990 Decision trees and decision-making. *IEEE Trans. Syst. Man Cybern.* **20**, 339–346. https://doi.org/10.1109/21.52545.

Quaghebeur, W., Torfs, E., De Baets, B. & Nopens, I. 2022 Hybrid differential equations: integrating mechanistic and data-driven techniques for modelling of water systems. *Water Research* **213**, 118166. https://doi.org/10.1016/j.watres.2022.118166

Raissi, M., Perdikaris, P. & Karniadakis, G. E. 2017 Machine learning of linear differential equations using Gaussian processes. *J. Comput. Phys.* **348**, 683–693.

Raissi, M., Perdikaris, P. & Karniadakis, G. E. 2019 Physics-informed neural networks: a deep learning framework for solving forward and inverse problems involving nonlinear partial differential equations. *J. Comput. Phys.* **378**, 686–707.

Razavi, S. 2021 Deep learning, explained: fundamentals, explainability, and bridgeability to process-based modelling. *Environ. Modell. Software* **144**, 105159.

Razavi, S., Tolson, B. A. & Burn, D. H. 2012 Review of surrogate modeling in water resources. *Water Resour. Res.* **48**. https://doi.org/10.1029/2011WR011527.

Refsgaard, J. C., van der Sluijs, J. P., Højberg, A. L. & Vanrolleghem, P. A. 2007 Uncertainty in the environmental modelling process – a framework and guidance. *Environ. Modell. Software* **22**, 1543–1556. https://doi.org/10.1016/j.envsoft.2007.02.004.

Regmi, P., Stewart, H., Amerlinck, Y., Arnell, M., García, P. J., Johnson, B., Maere, T., Miletić, I., Miller, M., Rieger, L., Samstag, R., Santoro, D., Schraa, O., Snowling, S., Takács, I., Torfs, E., van Loosdrecht, M. C. M., Vanrolleghem, P. A., Villez, K., Volcke, E. I. P., Weijers, S., Grau, P., Jimenez, J. & Rosso, D. 2019 The future of WRRF modelling – outlook and challenges. *Water Sci. Technol.* **79**, 3–14. https://doi.org/10.2166/wst.2018.498.

Reynaert, E., Hess, A. & Morgenroth, E. 2021 Making waves: why water reuse frameworks need to co-evolve with emerging small-scale technologies. *Water Res. X* **11**, 100094. https://doi.org/10.1016/j.wroa.2021.100094.

Rieger, L., Gillot, S., Langergraber, G., Ohtsuki, T., Shaw, A., Takacs, I. & Winkler, S. 2012 *Guidelines for Using Activated Sludge Models*. London, UK: IWA publishing.

Rodriguez-Roda, I. R., Sànchez-Marrè, M., Comas, J., Baeza, J., Colprim, J., Lafuente, J., Cortes, U. & Poch, M. 2002 A hybrid supervisory system to support WWTP operation: implementation and validation. *Water Sci. Technol.* **45**, 289–297. https://doi.org/10.2166/wst.2002.0608.

Romijn, R., Özkan, L., Weiland, S., Ludlage, J. & Marquardt, W. 2008 A grey-box modeling approach for the reduction of nonlinear systems. *J. Process Control* **18**, 906–914. https://doi.org/10.1016/j.jprocont.2008.06.007.

Russo, S., Lürig, M., Hao, W., Matthews, B. & Villez, K. 2020 Active learning for anomaly detection in environmental data. *Environ. Modell. Software* **134**, 104869. https://doi.org/10.1016/j.envsoft.2020.104869.

Sahoo, G. B., Ray, C. & Wade, H. F. 2005 Pesticide prediction in ground water in North Carolina domestic wells using artificial neural networks. *Ecol. Modell.* **183**, 29–46.

Samek, W., Wiegand, T. & Müller, K.-R. 2017 *Explainable Artificial Intelligence: Understanding, Visualizing and Interpreting Deep Learning Models*. ArXiv170808296 Cs Stat.

Samuelsson, O., Olsson, G., Lindblom, E., Björk, A. & Carlsson, B. 2021 Sensor bias impact on efficient aeration control during diurnal load variations. *Water Sci. Technol.* **83**, 1335–1346. https://doi.org/10.2166/wst.2021.031.

Schneider, M. Y., Carbajal, J. P., Furrer, V., Sterkele, B., Maurer, M. & Villez, K. 2019 Beyond signal quality: the value of unmaintained pH, dissolved oxygen, and oxidation-reduction potential sensors for remote performance monitoring of on-site sequencing batch reactors. *Water Res.* **161**, 639–651. https://doi.org/10.1016/j.watres.2019.06.007.

Schneider, M. Y., Furrer, V., Sprenger, E., Carbajal, J. P., Villez, K. & Maurer, M. 2020 Benchmarking soft sensors for remote monitoring of on-site wastewater treatment plants. *Environ. Sci. Technol.* **54**, 10840–10849. https://doi.org/10.1021/acs.est.9b07760.

Schubert, J., Simutis, R., Dors, M., Havlík, I. & Lübbert, A. 1994 Hybrid modelling of yeast production processes–combination of a priori knowledge on different levels of sophistication. *Chem. Eng. Technol. Ind. Chem.-Plant Equip.-Process Eng.-Biotechnol.* **17**, 10–20.

Sedlak, D. 2014 *Water 4.0: The Past, Present, and Future of the World's Most Vital Resource*. Yale University Press, New Haven, CT, USA.

Shapley, L. S. 2016 *17. A Value for n-person Games*. 307–18, Princeton University Press, 1953. Princeton. https://doi.org/10.1515/9781400881970-018.

Shi, S. & Xu, G. 2018 Novel performance prediction model of a biofilm system treating domestic wastewater based on stacked denoising auto-encoders deep learning network. *Chem. Eng. J.* **347**, 280–290. https://doi.org/10.1016/j.cej.2018.04.087.

Sharif Shourjeh, M., Kowal, P., Lu, X., Xie, L. & Drewnowski, J. 2021 Development of strategies for AOB and NOB competition supported by mathematical modeling in terms of successful deammonification implementation for energy-efficient WWTPs. *Processes* **9** (3), 562. https://doi.org/10.3390/pr9030562

Sin, G. & Al, R. 2021 Activated sludge models at the crossroad of artificial intelligence – a perspective on advancing process modeling. *Npj Clean Water* **4**, 1–7. https://doi.org/10.1038/s41545-021-00106-5.

Sitzenfrei, R., Hillebrand, S. & Rauch, W. 2017 Investigating the interactions of decentralized and centralized wastewater heat recovery systems. *Water Sci. Technol.* **75**, 1243–1250. https://doi.org/10.2166/wst.2016.598.

Stedmon, C. A., Seredyńska-Sobecka, B., Boe-Hansen, R., Le Tallec, N., Waul, C. K. & Arvin, E. 2011 A potential approach for monitoring drinking water quality from groundwater systems using organic matter fluorescence as an early warning for contamination events. *Water Res.* **45**, 6030–6038.

Su, H.-T., Bhat, N., Minderman, P. & McAvoy, T. 1993 Integrating neural networks with first principles models for dynamic modeling. In: *Dynamics and Control of Chemical Reactors, Distillation Columns and Batch Processes*. Elsevier, pp. 327–332. IFAC Symposia Series. Oxford: Pergamon, 1993. https://doi.org/10.1016/B978-0-08-041711-0.50054-4.

Su, Hong-Te, N. Bhat, P. A. Minderman, and T. J. McAvoy. "Integrating Neural Networks with First Principle Modles for Dynamic Modeling." In Dynamics and Control of Chemical Reactors, Distillation Columns and Batch Processes, edited by J. G. Balchen, 327–32. IFAC Symposia Series. Oxford: Pergamon, 1993. https://doi.org/10.1016/B978-0-08-041711-0.50054-4

Sutton, R. S. & Barto, A. G. 2018 *Reinforcement Learning: An Introduction*. Cambridge: MIT press.

Takács, I., Patry, G. G. & Nolasco, D. 1991 A dynamic model of the clarification-thickening process. *Water Res.* **25**, 1263–1271. https://doi.org/10.1016/0043-1354(91)90066-Y.

Talebizadeh, M., Belia, E. & Vanrolleghem, P. A. 2016 Influent generator for probabilistic modeling of nutrient removal wastewater treatment plants. *Environ. Modell. Software* **77**, 32–49. https://doi.org/10.1016/j.envsoft.2015.11.005.

Tesoriero, A. J., Gronberg, J. A., Juckem, P. F., Miller, M. P. & Austin, B. P. 2017 Predicting redox-sensitive contaminant concentrations in groundwater using random forest classification. *Water Resour. Res.* **53**, 7316–7331.

Therrien, J.-D., Nicolaï, N. & Vanrolleghem, P. A. 2020 A critical review of the data pipeline: how wastewater system operation flows from data to intelligence. *Water Sci. Technol.* **82**, 2613–2634. https://doi.org/10.2166/wst.2020.393.

Thompson, M. L. & Kramer, M. A. 1994 Modeling chemical processes using prior knowledge and neural networks. *AIChE J.* **40**, 1328–1340. https://doi.org/10.1002/aic.690400806.

Thürlimann, C. M., Udert, K. M., Morgenroth, E. & Villez, K. 2019 Stabilizing control of a urine nitrification process in the presence of sensor drift. *Water Res.* **165**, 114958. https://doi.org/10.1016/j.watres.2019.114958.

Torfs, E., Dutta, A. & Nopens, I. 2012 Investigating kernel structures for Ca-induced activated sludge aggregation using an inverse problem methodology. In: *4th Int. Conf. Popul. Balance Model.* Vol. 70, pp. 176–187. https://doi.org/10.1016/j.ces.2011.06.069.

Tsen, A. Y.-D., Jang, S. S., Wong, D. S. H. & Joseph, B. 1996 Predictive control of quality in batch polymerization using hybrid ANN models. *AIChE J.* **42**, 455–465.

Tulleken, H. J. 1993 Grey-box modelling and identification using physical knowledge and Bayesian techniques. *Automatica* **29**, 285–308.

Van Can, H. J., Hellinga, C., Luyben, K. C. A., Heijnen, J. J. & Te Braake, H. A. 1996 Strategy for dynamic process modeling based on neural networks in macroscopic balances. *AIChE J.* **42**, 3403–3418.

Van der Helm, A. W. C. & Rietveld, L. C. 2002 Modelling of drinking water treatment processes within the Stimela environment. *Water Sci. Technol. Water Supply* **2**, 87–93.

van Loosdrecht, M. C. M. & Brdjanovic, D. 2014 Anticipating the next century of wastewater treatment. *Science* **344**, 1452–1453. https://doi.org/10.1126/science.1255183.

Vanrolleghem, P. A. & Lee, D. S. 2003 On-line monitoring equipment for wastewater treatment processes: state of the art. *Water Sci. Technol.* **47**, 1–34. https://doi.org/10.2166/wst.2003.0074.

Vanrolleghem, P. A., Benedetti, L. & Meirlaen, J. 2005 Modelling and real-time control of the integrated urban wastewater system. *Vulnerability Water Qual. Intensiv. Dev. Urban Watersheds* **20**, 427–442. https://doi.org/10.1016/j.envsoft.2004.02.004.

Versteegh, J. F. M., Van Gaalen, F. W., Rietveld, L. C., Evers, E. G., Aldenberg, T. A. & Cleij, P. 2001 *TAPWAT: Definition Structure and Applications for Modelling Drinking Water Treatment.*

Villez, K., Ruiz, M., Sin, G., Colomer, J., Rosén, C. & Vanrolleghem, P. A. 2008 Combining multiway principal component analysis (MPCA) and clustering for efficient data mining of historical data sets of SBR processes. *Water Sci. Technol.* **57**, 1659. https://doi.org/10.2166/wst.2008.143.

Villez, K., Billeter, J. & Bonvin, D. 2019 Incremental parameter estimation under Rank-Deficient measurement conditions. *Processes.* https://doi.org/10.3390/pr7020075.

Villez, K., Vanrolleghem, P. A. & Corominas, L. 2020 A general-purpose method for Pareto optimal placement of flow rate and concentration sensors in networked systems – with application to wastewater treatment plants. *Comput. Chem. Eng.* **139**. https://doi.org/10.1016/j.compchemeng.2020.106880.

von Stosch, M., Oliveira, R., Peres, J. & Feyo de Azevedo, S. 2014 Hybrid semi-parametric modeling in process systems engineering: past, present and future. *Comput. Chem. Eng.* **60**, 86–101. https://doi.org/10.1016/j.compchemeng.2013.08.008.

Wade, M. 2004 *Process Monitoring and Knowledge Extraction in Wastewater Treatment Plants. PhD Thesis.*

Wade, M. J., Sánchez, A. & Katebi, M. R. 2005 On real-time control and process monitoring of wastewater treatment plants: real-time process monitoring. *Trans. Inst. Meas. Control* **27**, 173–193. London, UK. https://doi.org/10.1191/0142331205tm140oa.

Wade, M. J., Keedwell, E. C., Steyer, J.-P. & Ruano Garcia, M. V. 2021 *Making Water Smart, In Focus – Special Book Series.* IWA Publishing.

Wang, Z. & Georgakis, C. 2019 A dynamic response surface model for polymer grade transitions in industrial plants. *Ind. Eng. Chem. Res.* **58**, 11187–11198.

Wang, X., Zhang, J. & Babovic, V. 2016 Improving real-time forecasting of water quality indicators with combination of process-based models and data assimilation technique. *Ecol. Indic.* **66**, 428–439.

Wang, X., Kvaal, K. & Ratnaweera, H. 2019 Explicit and interpretable nonlinear soft sensor models for influent surveillance at a full-scale wastewater treatment plant. *J. Process Control* **77**, 1–6. https://doi.org/10.1016/j.jprocont.2019.03.005.

Wärff, C., Arnell, M., Sehlén, R. & Jeppsson, U. 2020 Modelling heat recovery potential from household wastewater. *Water Sci. Technol.* **81**, 1597–1605. https://doi.org/10.2166/wst.2020.103.

Wright, L. & Davidson, S. 2020 How to tell the difference between a model and a digital twin. *Advanced Modeling and Simulation in Engineering Sciences* **7** (1), 13. https://doi.org/10.1186/s40323-020-00147-4.

WOS 2021 *Web of Science Core Collection Help.*

Wu, G.-D. & Lo, S.-L. 2008 Predicting real-time coagulant dosage in water treatment by artificial neural networks and adaptive network-based fuzzy inference system. *Eng. Appl. Artif. Intell.* **21**, 1189–1195.

Yang, M., del Pozo, D. F., Torfs, E., Rehman, U., Yu, D. & Nopens, I. 2021 Numerical simulation on the effects of bubble size and internal structure on flow behavior in a DAF tank: a comparative study of CFD and CFD-PBM approach. *Chem. Eng. J. Adv.* **7**, 100131. https://doi.org/10.1016/j.ceja.2021.100131.

Zhu, J.-J. & Anderson, P. R. 2019 Performance evaluation of the ISMLR package for predicting the next day's influent wastewater flowrate at Kirie WRP. *Water Sci. Technol.* **80**, 695–706. https://doi.org/10.2166/wst.2019.309.

Zhu, J.-J., Segovia, J. & Anderson, P. R. 2015 Defining influent scenarios: application of cluster analysis to a water reclamation plant. *J. Environ. Eng.* **141**, 04015005. https://doi.org/10.1061/(ASCE)EE.1943-7870.0000934.

Zhu, J.-J., Kang, L. & Anderson, P. R. 2018 Predicting influent biochemical oxygen demand: balancing energy demand and risk management. *Water Res.* **128**, 304–313. https://doi.org/10.1016/j.watres.2017.10.053.

Zhu, J.-J., Dressel, W., Pacion, K. & Ren, Z. J. 2021 *ES&t* in the 21st century: a data-driven analysis of research topics, interconnections, and trends in the past 20 years. *Environ. Sci. Technol.* **55**, 3453–3464. https://doi.org/10.1021/acs.est.0c07551.

First received 10 January 2022; accepted in revised form 24 March 2022. Available online 6 April 2022

doi: 10.2166/wst.2022.107

The transition of WRRF models to digital twin applications

Elena Torfs [iD] a,b,*, Niels Nicolaï [iD] c,d, Saba Daneshgar [iD] a,b, John B. Copp [iD] e, Henri Haimi f, David Ikumi [iD] g, Bruce Johnson [iD] h, Benedek B. Plosz [iD] i, Spencer Snowling [iD] j, Lloyd R. Townley [iD] k,l, Borja Valverde-Pérez [iD] m, Peter A. Vanrolleghem [iD] c,d, Luca Vezzaro [iD] m and Ingmar Nopens [iD] a,b

a Biomath, Ghent University, Coupure links 653, 9000 Gent, Belgium
b Centre for Advanced Process Technology for Urban Resource recovery (CAPTURE), Frieda Saeysstraat 1, 9000 Gent, Belgium
c modelEAU, Université Laval, 1065 avenue de la Médecine, Québec G1 V 0A6, QC, Canada
d CentrEau, Québec Water Research Centre, 1065 avenue de la Médecine, Québec G1 V 0A6, QC, Canada
e Primodal, Inc., 122 Leland Street, Hamilton, Ontario L8S 3A4, Canada
f FCG Finnish Consulting Group Ltd, Osmontie 34, P.O. Box 950, FI-00601 Helsinki, Finland
g Water Research Group, Department of Civil Engineering, University of Cape Town, Rondebosch, 7700 Cape, South Africa
h Jacobs, 9191 Jamaica St, Englewood, CO 80112, USA
i Department of Chemical Engineering, University of Bath, Claverton Down, Bath, BA2 7AY, UK
j Hatch, Ltd, 2800 Speakman Dr, Mississauga, ON, Canada, L5 K 2R7
k Nanjing University Yixing Environmental Research Institute, 128 Hengtong Road, Yixing Jiangsu, 214200, China
l Nanjing Smart Technology Development Co. Ltd, Yanlord Landmark Building B Suite 706, Jiangxinzhou Jianye, Nanjing Jiangsu 210019, China
m Department of Environmental Engineering, Technical University of Denmark, Bygningstorvet, Building 115, 2800, Kongens Lyngby, Denmark
*Corresponding author. E-mail: elena.torfs@ugent.be

[iD] ET, 0000-0002-5629-6950; NN, 0000-0003-2794-2930; SD, 0000-0002-6407-1145; JBC, 0000-0002-5080-9997; DI, 0000-0001-9158-2525; BJ, 0000-0002-0012-6477; BBP, 0000-0002-7081-7038; SS, 0000-0001-8037-8983; LRT, 0000-0003-2923-628X; BV-P, 0000-0001-7255-1395; PAV, 0000-0003-1695-1313; LV, 0000-0001-6344-7131; IN, 0000-0001-6670-3700

ABSTRACT

Digital Twins (DTs) are on the rise as innovative, powerful technologies to harness the power of digitalisation in the WRRF sector. The lack of consensus and understanding when it comes to the definition, perceived benefits and technological needs of DTs is hampering their widespread development and application. Transitioning from traditional WRRF modelling practice into DT applications raises a number of important questions: When is a model's predictive power acceptable for a DT? Which modelling frameworks are most suited for DT applications? Which data structures are needed to efficiently feed data to a DT? How do we keep the DT up to date and relevant? Who will be the main users of DTs and how to get them involved? How do DTs push the water sector to evolve? This paper provides an overview of the state-of-the-art, challenges, good practices, development needs and transformative capacity of DTs for WRRF applications.

Key words: data management, hybrid models, interoperability, model predictive control, online optimisation, proactive management

HIGHLIGHTS

- A Digital Twin distinguishes itself from a simulation model by a continuous, automated data connection.
- Current DT projects in WRRFs focus on operational support and control.
- Combining mechanistic models and data-driven techniques into hybrid models can accelerate the adoption of DTs.
- Data and information models are key for the implementation and upscaling of DTs.
- WRRF staff should be included during the development stages of DTs.

GRAPHICAL ABSTRACT

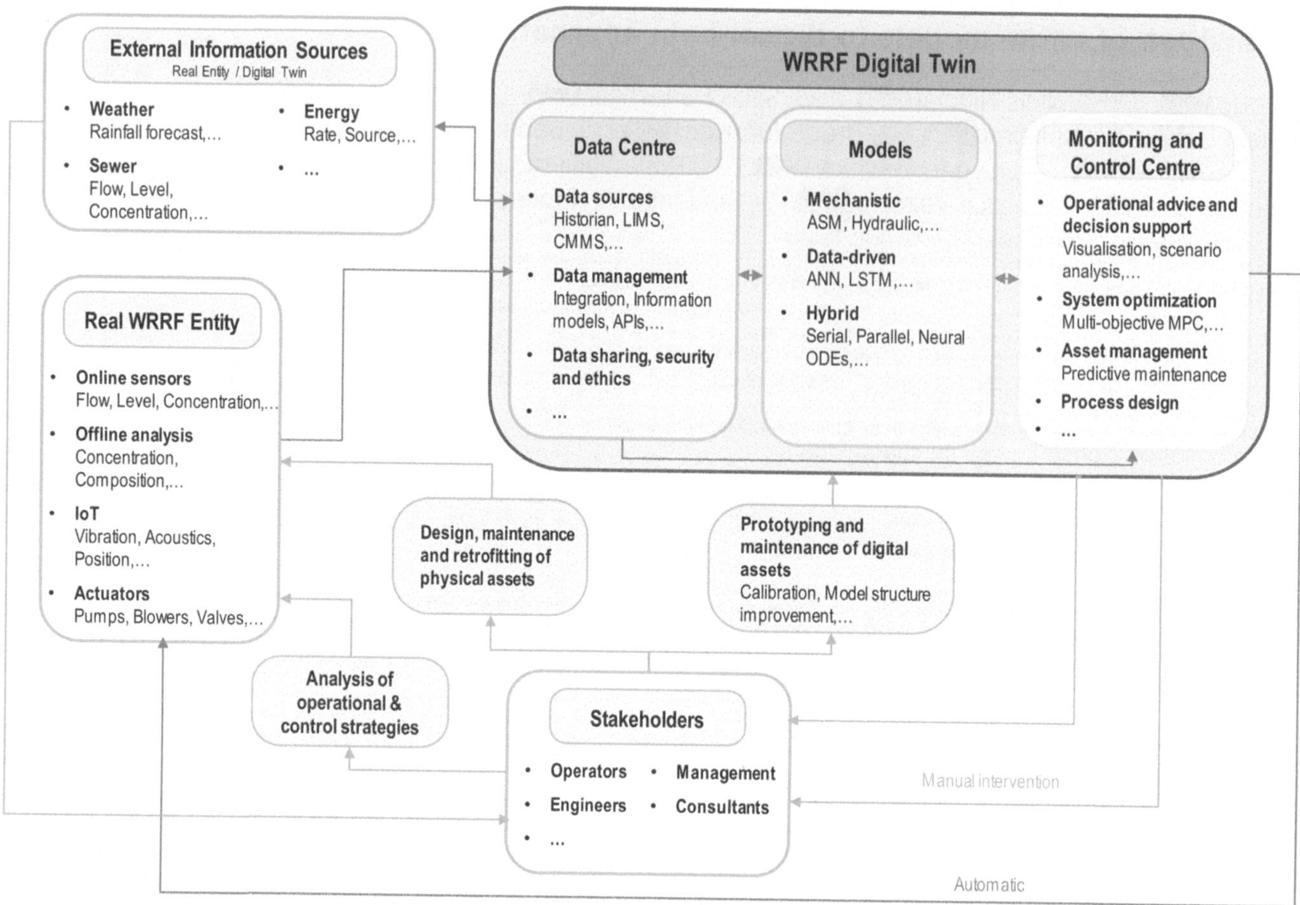

1. INTRODUCTION

The water industry is transforming and digitalisation is a key part of this transition (IWA & Xylem Inc 2019). However, digitalisation and its derivatives such as Digital Twins (DTs) seem to be buzzwords adopted from other industries and used without sufficient consensus and in-depth understanding of the challenges and development needs ahead. Moreover, many digitalisation initiatives seem to be skewed towards drinking water distribution applications (e.g. IWA's digital water program, the SWAN working group on Digital Twin applications etc.) with Water Resource Recovery Facility (WRRF) applications gaining less attention. This delayed uptake in the WRRF sector can be explained by the complex biological reactions that are governing typical treatment systems leading to specific challenges with data collection and model development.

The transition from conventional process models to DTs should be motivated by the plant's objective. As such, offline process models used for design and decision-making and DTs used for automated live process monitoring, control, and/or optimisation can be complementary. To achieve the transition to a DT, dramatic changes at both cultural and technical levels need to be put into practice. For example, the predictive power of conventional mechanistic modelling practices for WRRFs (e.g. ASM, ADM, and BSM) should be assessed and, if needed, existing model structures should be combined with data-driven techniques. The way water and wastewater utilities collect and store data can be optimised to have more relevant and consistent data. Finally, trust between the operational staff, leadership, and the DT will be needed. It is important to note that by looking more holistically, a WRRF DT can be integrated in a higher level decision support system. This multidisciplinary approach with inter-connection between various subsystems from different technical domains brings a whole new set of challenges and makes the need for some level of agreement on DT definitions and elements even more critical.

In order to realise the perceived benefits from the rapidly expanding concept of DTs, consensus is needed to provide clear definitions of what DT is. Currently, consensus is missing in the WRRF modelling community concerning the state-of-the-art, challenges, good practices, development needs and transformative capacity of DTs for WRRF applications. To overcome this barrier, a dedicated workshop was organised during the (virtual) 7th IWA/WEF Water Resource Recovery Modelling Seminar 2021. This paper summarises the results of this workshop and presents: (1) a definition of DTs in the WRRF sector based on key features that distinguish a DT from conventional simulation models, (2) guidelines on the transition of WRRF modelling efforts to successful application of DTs, (3) a comprehensive overview of technical and non-technical challenges and development needs for DTs, and (4) an outlook for the future of DTs as a transformative digital decision-making tool.

2. DIGITAL TWIN DEFINITION

The origin of using the DT terminology goes back to the early 2000s but it did not appear in official publications until the early 2010s where it was first introduced in the manufacturing industry (Glaessgen & Stargel 2012; Grieves 2014) as a new tool for the management of a product's lifecycle. Since then, the concept of DTs has spread throughout many different sectors including the water industry.

Despite the growing popularity of DT terminology in the domain of wastewater treatment, the distinction between DTs and conventional process modelling is not consistently defined, leading to the erroneous use of the term in many studies where models are applied for process design and decision-making. This inconsistency may lead to misunderstanding, contribute to the idea that DTs are merely hype and consequently may even slow down the adoption of DTs (Wright & Davidson 2020). Hence, there is an increasing need to define what differentiates DTs from conventional simulation models so that a common understanding on the topic can be maintained within our domain.

Many definitions of DTs can be found in the literature across different domains (Curl *et al.* 2019; Karmous-Edwards *et al.* 2019; Fuller *et al.* 2020; Wright & Davidson 2020). General consensus can be found in the idea that what distinguishes a DT from a conventional simulation model is its continuous **connection** with the physical twin (Jones *et al.* 2020). This connection should include the following features:

1. The idea of a twin inherently assumes that **there exists a physical counterpart**. Hence, a DT should be connected to a real entity (e.g. equipment, process or product) even though its development process can start as soon as the real entity is in its conceptualisation stage. This is especially relevant in discussions on the potential of DTs for design (greenfield facilities, retrofitting existing reactors, extending existing plants with new treatment lines).
2. A DT should have an **automated, live data connection to the real entity**. This connection is preferably bi-directional such that insight generated by the virtual twin is fed back to the real entity. Whereas the data connection from the real entity to the DT should be automated, the virtual-to-real entity connection may be either automated (direct input from the DT into SCADA) or manually performed using humans in the loop.
3. Twins should evolve together. Hence, the DT should include the means to **dynamically update** or adjust the models based on relevant data to maintain an accurate description of the real entity as it evolves over time. A DT thus becomes a continuously updated knowledge repository that is automatically kept relevant and thus provides a much higher level of usefulness as compared to a conventional simulation model executed independently.

In the context of wastewater treatment and resource recovery, the definition and application of DTs has been discussed in a recent white paper by the IWA specialist group on modelling and integrated assessment (IWA 2021a) as well as during a dedicated workshop at the WRRmod2021 conference. Although there is general agreement on the features listed above, there are still unresolved questions particularly with respect to the aspect of real-time connectivity. A real-time data connection is an ill-defined concept because the time frame is arbitrary. Are data from 1 hour or 1 day ago still considered real-time? Therefore, the authors propose that the time horizon for model updates and simulation should be defined by the DT objectives (e.g. instrument failure detection, real-time control) and the relevant system dynamics (aeration vs. sludge age). Hence, the automated data-feed and dynamic model updating are what defines a DT whereas the time horizon over which this is performed is flexible.

An overview of the components of a DT and its connections to the real entity and relevant stakeholders is provided in Figure 1.

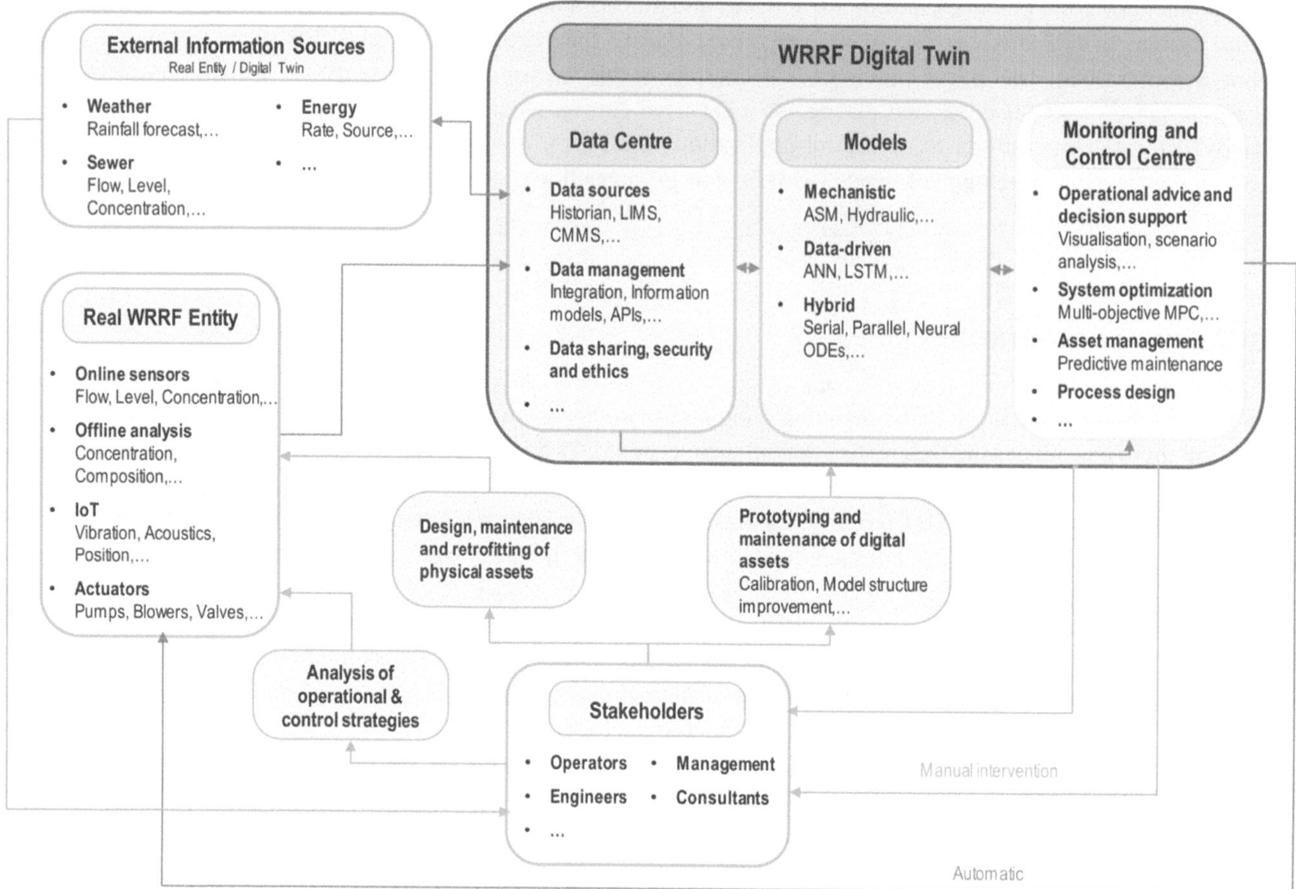

Figure 1 | Schematic representation of a Digital Twin and its interaction with the physical system, control systems and stakeholders.

3. DIGITAL TWIN APPLICATIONS IN WRRFS

In general, the application of DTs supports operations and management staff in making data both time-relevant and actionable in a way that traditional data analysis and modelling cannot. Based on data from the physical asset or system, a DT unlocks value principally by supporting improved decision-making, which creates the opportunity for positive feedback into the physical twin (Bolton *et al.* 2018).

As such, in WRRFs, DTs can support the transition towards proactive management, whereby different processes and assets can be operated and maintained, to mitigate disturbances and other issues before they have undue adverse impacts on performance (Karmous-Edwards *et al.* 2019). As a result, there is significant potential for economic savings (e.g. energy optimisation) and more effective protection of the environment (e.g. better nutrient removal and recovery, reduction in GHG emissions). With WRRFs contributing approximately 1–3% of total global energy consumption (Mamais *et al.* 2015) and about 2–3% of total global GHG emissions (Maktabifard *et al.* 2019), significant gains can be foreseen from real-time optimisation and advanced control supported by DTs. Moreover, DTs can accelerate the transition to a circular water economy by tailoring effluent quality towards water reuse purposes.

DT applications include (but are not limited to): (i) data-driven decision support for the selection of different operational strategies and operator training (Johnson *et al.* 2021); (ii) online (multi-objective) system optimisation (e.g. model-predictive control) for energy or resource savings or compliance management (e.g. to minimise carbon footprint) (Stentoft *et al.* 2020, 2021); (iii) failure analysis (Jain *et al.* 2020); (iv) asset management and predictive maintenance (Bartos & Kerkez 2021); (v) investment planning (Ruohomäki *et al.* 2018); and (vi) decision support for policy making (Poch *et al.* 2020). Although the potential for application of DTs is far-reaching, only a limited number of successful implementations are documented in the literature for WRRF systems and they mainly focus on operational decision support and online system optimisation through advanced control. In the list above, for example, only Johnson *et al.* (2021) and Stentoft *et al.* (2020, 2021) describe applications in the WRRF sector, whereas other examples are taken from related domains such as water network management, smart cities, and the manufacturing industry. Some examples of application of DTs in the WRRF sector for operational decision making and online optimisation are described in more detail below.

3.1. Digital twins for operational advice

A DT was developed for the Singapore PUB Changi Water Reclamation Plant (WRP) (Johnson *et al.* 2021). This DT includes the whole plant process, hydraulics, and controls and automatically accepts over 1200 data streams from both SCADA and Laboratory Information Management systems. It is implemented on a dedicated server system located at the Changi WRP accepting data from a replicated historian system. It has been implemented as a hybrid model that uses both mechanistic and data-driven models, and includes some level of automatic calibration through the use of both the SCADA and laboratory data connections. The Changi WRP DT includes current status evaluations of measured versus DT results, automated scenario analysis (i.e. what happens if secondary settling tanks or bioreactors are out of service?), and a 5-day hourly prediction (a wastewater 'weather' forecast) of the plant performance. Operations staff are free to use the information provided by the DT to help them transition to a more proactive operational mode. It has been implemented in the advisory-only mode at this point to provide operations staff time to both improve accuracy and develop the needed trust prior to proving its control authority. The DT can also be used for operator training as it allows the operators to 'operate' the facility virtually in a safe environment under conditions that are rarely observed. These conditions include equipment failures, extreme high or low flows, emergency shutdowns and system restarts.

3.2. Digital twins for model predictive control

Model predictive control (MPC) allows for online optimisation of WRRFs. Stentoft *et al.* (2020) presented an MPC for integrated control of a pumping station in the catchment of Kolding WRRF (Denmark). The aim of the controller was to minimise energy usage while respecting discharge limits. The DT consisted of a catchment model (including both the sewer line and dry weather inlet flow forecast) as well as an effluent quality model, both being empirical models calibrated on an hourly basis. The effluent quality model uses measurements of ammonia and nitrate concentrations at 2 minute intervals to approximate the effluent TN. First, the data are resampled to hourly values and then, model parameters are estimated using maximum likelihood estimation and a Kalman filter. The catchment model was, however, calibrated offline and predicts the flow rate based on the time of the day under dry weather conditions. The DT is implemented as an extra module of the control system. The example falls within the scope of a DT, as the models are dynamically updated with data from the physical twin, while the DT can also communicate with the physical counterpart to improve its performance. The MPC optimises the flow rate of the pumping station every hour with a prediction horizon of 24 hours. The controller was successfully validated during two operational periods over a total of 7 days. Currently, the developer is working on an extension of the MPC to optimise WRRFs based on competing operational objectives, such as cost, effluent quality, and carbon footprint (Stentoft *et al.* 2021). To that end, new data-driven models need to be developed and validated, as it is important that the DT adapts to the new optimisation needs.

3.3. Digital twins for design

The potential application of DTs for design of WRRFs is a topic of much discussion as automated data coupling is often lacking when an entity is in its design phase. However, in the manufacturing industry several applications of DTs for product design can be found (Wright & Davidson 2020): so-called prototyping DTs are fed with data from existing objects or products and used to update the design of new versions, which may or may not reach the commercial phase, but generate new data to update the DT. Within the WRRF sector, the main interest is currently on DTs for operational purposes, as current practice for model-based design does not include any automated data coupling (Pedersen *et al.* 2021). This, however, does not mean that a model developed in a WRRF's design phase cannot transition into an operational DT as the reactor or facility goes into operation. A recent case study that investigates this transition between design and operation is the work performed by Alex *et al.* (2020), which illustrates how a detailed process model of a plant including its control system and actuators is used for virtual commissioning of the automation system. This process model developed during the planning and design phase of the automation concept can continue its lifecycle as an operational DT once the automation system is connected to the real entity. A similar example is the use of CFD models for design and DT development. Whereas CFD models are not yet directly used in DT applications for WRRFs, a detailed CFD model developed in the design phase can serve as an important information source for the development of an operational DT through the derivation of compartmental models (Le Moullec *et al.* 2011). Finally, DTs of existing systems can be powerful tools for evaluating potential process design changes because, by definition, the DT represents an accurate and relevant description of the real entity's current state.

4. MODELLING TOOLS AND PREDICTIVE POWER

A DT has a mathematical model at its core. In principle, a DT can use any type of model that is a sufficiently accurate representation of its physical counterpart (mechanistic or first principles models; fully data-driven models; or a combination of both, so-called hybrid models). In practice, the choice of model structure will be influenced by a number of factors that are specifically related to the application and specific objectives of the DT as this will determine the timescale over which a DT needs to be evaluated and updated. Given the dominant dynamics of most governing processes in WRRFs, this will likely be in the range of minutes, hours or days for operational purposes. However, for applications such as predictive maintenance or investment planning, the timescale can be much larger (Wright & Davidson 2020).

In an ideal world where computational power is not an issue and all processes in WRRFs are perfectly understood, DTs would make use of so-called mechanistic or first principles models. Mechanistic models describe the system in a fixed, structured way derived from underlying physical, chemical, and biological mechanisms, often represented by balance equations of some quantities which change in time and space. The parameters of a mechanistic model have a physical meaning in the system. In general, mechanistic models have a relatively high computational power requirement making them less suitable for high-frequency DT applications where model updates are needed in real-time (for example for MPC applications) (Pantelides & Renfro 2013; Pedersen *et al.* 2021). For DT applications with a longer actionable time-horizon such as maintenance or investment planning, mechanistic models are highly relevant provided that their predictive power is sufficiently high.

At the other end of the spectrum, data-driven or so-called black box models describe the system purely based on information extracted out of the process data. The difference between pure black-box models (for example neural networks) and empirical relations derived from data (such as Monod kinetics or settling velocity functions) lies in their unstructured nature – the number and nature of the parameters are flexible and not fixed in advance by knowledge (von Stosch *et al.* 2014). These models do not incorporate any information on the underlying process dynamics and are therefore only reliable within the region of input parameter space from which the data used to construct the model was taken. Extrapolation of data-driven models beyond the operating space for which they were developed is a dangerous approach. Therefore, compared to mechanistic models, data-driven models need larger datasets to accurately predict a range of operational conditions. Moreover, their lack of interpretability is another issue for their use in DT applications. However, data-driven models are much less computationally expensive, making them very interesting for real-time DT applications such as data-driven soft sensor development for real-time monitoring (Haimi *et al.* 2013) and model predictive control for online optimisation (Stentoft *et al.* 2018). Historically, WRRF modellers are accustomed to developing and using mechanistic models whereas data-driven models are not yet widely used for these systems. This is related to the lack of expertise in data-driven techniques within the WRRF modelling community and the absence of sufficient trust in black-box modelling performance. It is also linked to the inherent nature of WRRF processes which are non-stationary and have significant temporal dependency which requires frequent retraining of the data-driven model. A comprehensive overview of barriers to the application of Big Data analytics in the domain of WRRFs as well as further examples of applications of data-driven models for real-time monitoring, fault detection, and control in WRRFs can be found in Newhart *et al.* (2019).

Two possible solutions exist to overcome the disadvantages of both mechanistic and data-driven models. A first very promising approach is the combination of both modelling paradigms into hybrid models (Lee *et al.* 2005; Quaghebeur *et al.* 2022). This creates a modelling paradigm that includes the best of both worlds: a mechanistic backbone incorporating relevant process knowledge and thus providing interpretability and extrapolation capabilities, as well as a data-driven part that augments the overall model's predictive power by including information on lesser-known subprocesses at reduced computational cost. An overview of different hybrid model configurations and applications for advanced control and online optimisation can be found in von Stosch *et al.* (2014).

A second solution to overcome the high computational demand of mechanistic models is the development of surrogate models. Surrogate models are simplified approximations of a more complex model of the system that statistically relate the input and output of the system. They are considerably useful when the underlying relationship between input and output of the system is unknown or in the present context computationally expensive to evaluate. Here, a highly accurate mechanistic model is used to generate a set of (simulated) data within the relevant operating space which can subsequently be used to train a data-driven model (Chinesta *et al.* 2020). Depending on the definition, the creation of surrogate models can also be seen as hybrid modelling practice.

Regardless of the selected model to be used in a DT of a WRRF, a challenging aspect to include in a DT is the means to **dynamically update** or adjust the models based on relevant data to maintain an accurate description of the real entity as it evolves over time. Therefore, the models need to be validated frequently and if necessary recalibrated or retrained. While this can be done in a manual way, automatic validation/(re)calibration of the models is one of the key aspects of DT. This is a crucial step in developing a DT so that it can deal with the known uncertainties in WRRF modelling without the need to interrupt the operation of the DT. There are some important aspects to consider for the implementation of such a validation/(re)calibration procedure. For example, choosing the frequency for the validation/(re)calibration could be dependent on the source and frequency of the available data (e.g. laboratory data versus online sensor data), and the nature of the process and parameters to be calibrated (e.g. settling vs biological processes). Typical parameter estimation methods for model (re)calibration are optimisation methods such as recursive weighted least square methods and Bayesian estimation using a Markov chain Monte Carlo approach. However, these techniques can be time-consuming and computationally expensive. A data-driven approach, like artificial neural networks, could be an appropriate choice for model (re)calibration in real-time (Samad & Mathur 1992; de Almeida Martins *et al.* 2021). However, the performance of these models is highly dependent on the quality of the training datasets, and they may need to be retrained from time to time should completely new scenarios appear in the data patterns. Existing model calibration protocols that are recognised as part of the Good Modelling Practices (GMP) do not account for the new challenge of automatic online calibration (Rieger *et al.* 2012). This does not necessarily mean that the traditional protocols are obsolete, but rather emphasises the need for changes or adaptations based on the specific requirements of a DT compared to traditional modelling practices.

Alternatively, state estimators such as extended Kalman filters and other data assimilation methods can be included in the DT to match the model predictions to the most recent observations. In this way the model can continuously account for measurement inaccuracies and slowly changing process dynamics (Pantelides & Renfro 2013; Afshari *et al.* 2017). However, the number of states to be corrected by state estimation methods cannot exceed the number of measurements. Hence, a subset of uncertain/varying states should be selected on which online model correction through state estimators can be applied (Patwardhan *et al.* 2007).

In summary, state estimators allow for continuous model correction to account for small inaccuracies and slowly changing dynamics whereas major changes in the operating or influent space (for example a shift in microbial dynamics) require recalibration of the DT model.

5. DATA MANAGEMENT IN WRRFS

As stated in Section 2, the distinction between a DT and a conventional simulation model is primarily determined by its automated and continued connection to the real entity. This connection is achieved by (bi-directional) data transfer. Indeed, if a DT is to consistently represent the structure, state, and/or behavior of its physical counterpart, then it needs to be provided with operational data in various forms. Different categories of data usage can be identified for a DT, e.g. to calibrate and train mechanistic and data-driven models, to supply models with input data for simulation, to evaluate the state of the dynamic subsystems, to feed soft sensors, and/or to provide the necessary inputs for automated closed-loop control strategies. Ideally, both the objectives of the DT as well as the means towards its objectives, i.e. the models and control strategies, will determine which data are required. In practice, however, there exist technological and monetary constraints that limit data collection and management at WRRFs, which means data access will ultimately dictate how the DT can optimally exploit the available data. Even with limited data, a DT can add value by providing context and making the data actionable.

Fortunately, the amount and reliability of data that is being collected by WRRFs is on the rise (Olsson *et al.* 2014). New, more reliable sensor technologies (e.g. liquid phase N_2O (Fenu *et al.* 2020)) combined with faster computations are making the storage of frequent and high-resolution water quantity and quality data more affordable. For a DT focused on the operation and maintenance stage of the WRRF lifecycle, this steady increase of real-time process signals offers opportunities to determine and deliver a proactive course of action for WRRFs (GWRC 2021).

While WRRF operators and engineers are focused on the dynamic signals that relate to plant performance, at the same time, other organisational activities that support WRRFs (e.g. equipment and infrastructure maintenance, finance, resource planning, compliance, research, and development, etc.) are also generating significant amounts of digitised data. Although the water sector is lagging behind other sectors in adopting these digital technologies (TWI2050 2019), the information gathered from digital workflows provides the context for interpretation of WRRF measurements and states. For example, dynamic time series data stored in a process historian can be routinely annotated with the rich but static and unstructured contextual data often found in operational logbooks, and computerised maintenance management systems. In this way, a necessary base for fault detection is made available. In contrast to process measurements, those metadata that relate to process equipment, infrastructure, location, maintenance history, quality, purpose, range etc., typically do not need to be updated at high frequency (IWA 2021b). However, they still need to be curated actively if reliable conclusions are to be extracted from a DT. This means that high levels of workflow organisation and automation are needed to ensure that new information generated by one party becomes rapidly available to all WRRF stakeholders. Data curation also requires that external contractors' non-editable contextual document handover (e.g. PDF files of P&IDs, screenshots of CAD models, protected automation logic, etc.) be replaced with data handover (e.g. P&IDs, CAD models and automation programs in file formats native to the design software) that seamlessly integrate into a DT (Brendelberger *et al.* 2019; Wiedau *et al.* 2019).

To understand the challenges related to data and their usage in the context of a DT for the water and wastewater industries, abstract representations of data flows can be useful. Therrien *et al.* (2020) describe the concept of a data pipeline that illustrates how raw data are transformed into intelligent action. A crucial weak spot in this flow is the need for qualitative data upon which operators and engineers can make reliable decisions, and mathematical models can be fine-tuned. The struggle for high quality data is particularly applicable to WRRFs, as these data are naturally dependent on various water quality sensors that are notorious for low veracity when exposed to the harsh wastewater medium combined with inadequate maintenance (Therrien *et al.* 2020). The COVID-19 pandemic has highlighted this problem. As on-site staff were reduced and selective maintenance programs were introduced, the quality of online sensor data collected in about one-third of treatment plants around the world decreased considerably (Rahman *et al.* 2021).

The relation between DTs and data quality can be compared to the causality dilemma of the chicken or the egg. While the quality of DT predictions is inherently influenced by the quality of the data input, the DT will also improve the quality of collected data by providing the means for automatic and live data checks and contextualization. Thorough data analysis and reconciliation to detect, isolate, identify and correct measurement faults (Olsson & Newell 1999) should therefore be included early in the conception of the DT. Given the short-term operational objectives, data reconciliation methods should be sufficiently automated to allow for quick decisions and rapid action (De Mulder *et al.* 2018). While mechanistic models such as mass balances can assist in data reconciliation (Spindler & Vanrolleghem 2012; Le *et al.* 2018), the bulk of this process will likely be performed by generically applicable data-driven methods (Corominas *et al.* 2018; Newhart *et al.* 2019). Using data-driven approaches means that periodic re-training will be needed to compensate for the time-varying nature of most environmental time series. This is preferably achieved using automatically adapting fault detection techniques (Haimi *et al.* 2013). This will lead to a need for sufficient training data so that domain experts are not overloaded with routine data labelling tasks required for supervisory model training. Instead, clever methods should be devised that exploit uncertainty in the detection to iteratively improve future model prediction performance while limiting the need for human input (Russo *et al.* 2020).

Collecting data without a proper management strategy can result in underutilised data sets that lack the necessary structure for users and/or automated tools to efficiently interact with the data (IWA 2021b). An important piece of any DT's application is therefore consistent with the vision put forward by the Findable Accessible Interoperable Reusable (FAIR) guiding principles for scientific data management and stewardship (Wilkinson *et al.* 2016). That is, data should be FAIR within the boundaries and permissions needed for the DT to operate. A key aspect of the FAIR principles is the adoption of open interfaces and protocols for authorised data access, such that vendor lock-in for both software and hardware is abolished. This philosophy for open data has recently also been adopted by some WRRF simulation software providers, which nowadays provide built-in support for open communication protocols used in industrial automation (e.g. OPC UA), connectivity to commercial cloud computing infrastructure (e.g. MS Azure), as well as application programming interfaces (API) for open-source scripting languages (e.g. Python) to name a few. The latter is especially important in the context of hybrid modelling. APIs that allow for easy coupling of data-driven modelling techniques and tools to existing simulators could facilitate the transition of the WRRF modelling community from the purely mechanistic paradigm to applications with hybrid models.

A secondary benefit that stems from FAIR data is the opportunity for data monetisation. Clean data streams can be offered via a controlled marketplace, which developers can utilise to develop new applications and business models (FIWARE 2021). Subscription-based decentralised networks, based on blockchain technology, are already available for secure data sharing while guaranteeing fair profits to its owners (cf. Streamr, http://streamr.network). The opportunity for new revenue streams for WRRF utilities in a data economy may also provide additional incentive for the adoption of FAIR data principles. Moreover, opening data to the research community or the general public can help in advancing data-driven research breakthroughs by introducing solutions from entirely different fields, while also creating public awareness (Bonabeau 2009; Quay *et al.* 2021).

Underlying all the mathematical modelling and data management lies the IT software architecture and hardware infrastructure needed to support the DT. Here, choices will have to be made based on maintenance needs, user interaction, and cybersecurity of the DT. This includes choosing between on-premises software or cloud solutions, open-source or closed-source software, and off-the-shelf or custom software applications. With respect to data management, the DT will probably need to query various operational databases of a WRRF. This can include process historians, laboratory information systems (LIMS), and computerised maintenance management systems (CMMS), but also databases that are related to business activities higher up in the organisational structure like those used for finance and human resources. It is also not beyond the realm of possibility that the DT may need to access data sources external to the organisation like those providing multi-day weather forecasts, energy prices or smart city data brokers. At the same time, new types of data will have to be stored and version controlled by the DT. These include simulation results, calibration/training data sets, model performance metrics, historic parameter values, input data for the modelling of future scenarios, user interaction, and clickstream data, etc. It should thus be clear that sophisticated data architectures and computational workflows, composed of structured and unstructured data import, storage, and processing, will be needed to serve the DT. According to Fuller *et al.* (2020), the cost of installing and running such infrastructure is likely to be the biggest challenge for any DT project. Moreover, with the advent of increased internet connectivity and the introduction of IoT tools, questions need to be addressed around data sharing, cybersecurity, and ethics (GWRC 2021).

The successful adoption of open data models, FAIR data principles and powerful data architectures is of course influenced by the culture of the organisation but also of the regulatory frameworks surrounding water management. Whereas data standardisation, openness and interoperability are high on the development agenda in Europe and North America, the same cannot be said for many other regions world-wide. In such countries or regions, the first step towards DT applications should naturally be a paradigm shift from data acquisition based solely on regulatory compliance to being driven by the need for holistic and sustainable multi-criteria water management practices.

6. DIGITAL TWINS BEYOND THE FENCE OF WRRFS

DTs may be developed for a range of purposes and operate at different scales. A DT could be as simple as a component or model-based controller for a specific task. However, a DT can also be scaled up, to be plant-wide, then to the integrated urban watershed, to city-wide, and beyond. The scale on which a DT is developed and deployed is determined by the objectives. WRRF DTs could benefit from integrating data sources from outside their fence such as weather forecasts (Heinonen *et al.* 2013), energy tariffing (Aymerich *et al.* 2015), data on the ecological status of receiving waters (Muschalla 2008) etc., for decision support. Such novel sources of information provide new opportunities to optimise and manage WRRFs from a holistic perspective rather than a domain-specific one. DTs of WRRFs can not only improve treatment plant performance but can also be used to serve societal benefits. For example, the successful implementation of reuse strategies for wastewater effluent will depend on the specific composition of waste streams but also on the needs and opportunities in the surrounding urban/industrial/natural landscape, e.g. reuse of wastewater effluent for irrigation in agriculture (Neto *et al.* 2021), use in industry, aquifer replenishing for indirect potable reuse (Bullard *et al.* 2019) etc. Looking even further outside of the scope of the water sector, DTs can be used to predict the impact of a new or retrofitted wastewater treatment plant on energy systems, e.g. potential power demand for pumping and aeration, supply from onsite renewable power generation, waste heat recovery potential from treated wastewater etc.

A high-level DT would allow interdependencies across sectors to be understood in a way that organisation level or sector-based DTs could not satisfy. However, for the water sector, we are not yet there: bringing modelling tools together over different scales to address high-level technological or societal challenges is still rarely achieved and few DTs at present are connected or share data across organisations, sectors or geographies. Lack of interoperability is a key constraint. Achieving a high level DT requires effective communication between different subsystems to allow automatic reasoning and optimisation in support of decision making. Hence, it is not a matter of developing a single all-encompassing model of a region but rather a matter of building the foundation for an ecosystem of interconnected DTs where interoperability and transparency are key through common and automated data and context information models (Bolton *et al.* 2018).

Exchange of data from potentially different domains, each characterised by a specific jargon, leads to the need for standardised data structuring and communication conventions. Several initiatives are emerging aimed to develop semantic information models or ontologies consisting of controlled vocabularies, relationships, constraints, and rules to provide organised, stable, and shareable data for specific domain contexts which will significantly increase the accessibility and transferability of data. Examples can be found in the domains of geospatial sciences (OGC 2021), public health environmental surveillance (PHES-ODM 2021), chemical process industry (DEXPI 2021), production systems engineering (AutomationML 2021), building information management (ISO 2018). Within the specific context of smart cities, the Open and Agile Smart Cities initiative (https://oascities.org/) has defined minimal interoperability mechanisms (MIMS) including common data models (DTDL 2021; FIWARE 2021) to ensure cross-sectorial communication of data and models. The goal of each of these initiatives is to capture and map complex relationships that exist in the physical world and translate them into the digital data world using a set of concepts and categories defined by the domain ontology.

Some first examples of high-level DTs based on common data and information models can be found in literature of other domains: the application of a decision support system for real-time air quality control in the city of Singapore by coupling weather predictions with combustion processes and city topology models (Farazi *et al.* 2020) or even the construction of a national DT for energy management in the UK (Akroyd *et al.* 2021).

7. THE HUMAN FACTOR IN A DIGITAL WORLD

Effective use of a DT will require a significant cultural change across an organisation. WRRFs that are mostly run from a reactive stance will shift to proactive understanding based on a stream of (near) real-time information. This demands that the entire workforce understands the value of the data they generate or maintain with respect to other activities that exist along an organisation's data pipeline.

In most ongoing DT projects in the WRRF sector, the primary user of the DT will typically be an operations staff member who is accustomed to having to make day-to-day decisions, based on sometimes incomplete and/or delayed information (e.g. laboratory tests). A DT opens a whole new spectrum of potentially relevant information. Hence, it is important that the DT is designed taking into account the needs of frontline staff, i.e. to present the right information in the right format for operational decision-making. If it can make their life easier in some significant way, without adding more work, WRRF operators are likely to adopt this technology enthusiastically. However, it is important to understand that frontline staff have varying degrees of digital literacy, i.e. familiarity and comfort with digital tools and workflows. Moreover, users of any DT will need to gain trust in the suggestions provided by the DT, especially in the case when these suggestions are directly used to control the plant's operation.

It is up to each organisation's leaders to facilitate this novel synergy between human and DT. Involving frontline staff in the development of the DT content and design and giving staff an ownership stake in the success of the technology is critical. This includes education and training, such that it is very clear what the DT can and cannot do. It is not uncommon for a layman to think that a DT can do everything (Fuller *et al.* 2020), but when something does not match these expectations, the entire DT is branded a failure, a situation that is hard for an organisation to recover from. Therefore, it is also advised that any new DT goes through a proving stage to give all staff confidence in its suggestions and/or actions.

Eerikäinen *et al.* (2020) interviewed employees of WRRFs and other related stakeholders about their expectations of new data tools. It was found that they were expecting a next generation of tools in the near future – tools that combine the competencies of both automation/software providers and wastewater process experts with a thorough understanding of process behaviour. It is clear that a DT makes use of relatively novel technologies stemming from a wide range of technical information and engineering domains. Having in-house access to all this expertise is highly unlikely for most water and wastewater utilities. New hiring policies and lifelong learning initiatives can help utilities overcome some of the knowledge and skill gaps but implementation of the DT will most likely be performed to some degree in partnership with consultancy firms and academic partners. Moreover, some of the data interpretation within the DT may also be outsourced. Subscription platforms for the external, certified, evaluation of specific measurements already exist to date (e.g. vibration and acoustic signal analysis for condition monitoring of critical assets) (cf. Zensor, https://www.zensor.be/). By using IoT devices, data are sent directly to cloud applications of service providers for specialised analysis and subsequent presentation to non-experts.

Eventually, the DT could tap into this external knowledge base to provide even more relevant and accurate predictions. While depending on external parties may provide answers, it also comes at additional costs. A workaround that is regarded as a key enabler for DTs are so-called low-code and no-code application development platforms (Michael & Wortmann 2021). By providing modular preconfigured building blocks with standardised communication interfaces in a user friendly, often graphical environment, the complex computational workflows at the basis of the DT can become manageable by the non-experts themselves which reinforces their degree of ownership of and trust in the DT technology.

Lastly, it is important to recognise the potential of young water professionals in the digital transformation of the water sector, including the introduction of DTs. With estimates on water workforce retirement in the next 10 years ranging from 30 to 50% in the USA, a new wave of millennials is expected to reinforce the sector (Dickerson & Butler 2018). Raised in an online world, these digital natives have (un)consciously created a distinctive way to look for information and solve challenges faced on the modern work floor and beyond (IWA 2021b). Of course, the continued beneficial use of a DT depends on accurate documentation, a maintenance program and succession plans should a company's champions leave. However, the fact that the DT itself ideally acts as a high fidelity knowledge repository composed of models and data on the WRRF past, present, and future can contribute significantly in the transfer of knowledge from experts to inexperienced newcomers. Hence the DT could also help mitigate knowledge loss as a result of the ageing workforce (Kadiyala & Macintosh 2018).

8. CONCLUSIONS

In many ways, the advent of DTs has the potential to entirely change the way utilities operate and design their facilities. The existence of a live, automated data connection between the DT and the real entity ensures that data is actionable and knowledge is up to date. As such, utilities can transition from reactive to proactive and holistic management. However, accomplishing such aspirational goals requires a combination of technological advances and a clear buy-in (through clear social and financial benefits) from relevant stakeholders involved.

Despite growing interest, successful full-scale applications in the WRRF sector are still rare. The current focus of DT development in the WRRF sector lies mostly with online operational support and real-time advanced control. The potential of other applications that are well documented in other sectors, such as asset management, retrofitting, etc. should however not be overlooked.

The predictive power of the model used in a DT as well as its ability to adapt and respond to new scenarios and disturbances is one of the key aspects required to build trust among operational staff and utility managers. This concerns not only the application of DTs but is also important for keeping them alive and relevant in the long-term. Hence, developing a model that is highly predictive with a robust validation/(re)calibration protocol on top is critical. Hybrid models can be important tools in boosting predictive power by balancing mechanistic models with data-driven techniques. Their application and development for WRRF processes can encourage and amplify the successful implementation of DTs. Moreover, the data connection of the DT to its physical counterpart is crucial to ensure a continuous feed of high-quality data to the modelling core of the DT. A proper data architecture and data management strategy as well as automated data analysis and reconciliation are essential for any DT project. DT projects should hereby follow the philosophy put forward by the Findable Accessible Interoperable Reusable (FAIR) guiding principles for data management.

The development of DTs opens up an enormous potential for optimisation outside the fence of a WRRF. Open data models and standards allow connection of WRRF DTs to other smart city data sources. Thus, to foster and stimulate truly holistic decision making in the water sector, any DT should be considered as part of a modular inter-connected structure using common and automated data and context information models supporting minimum interoperability mechanisms (MIMS) that allow automated reasoning over their combined structure.

Finally, the expectations of the DT end users should match the capabilities and limitations of the models. No technology is useful unless people want to use it and trust it. This means different things to different stakeholders. The organisational structure around the DT should account for these differences. This emphasises the importance of including WRRF staff during the development of the DT.

ACKNOWLEDGMENTS

This position paper gives an overview of the discussion that took place at the similarly titled workshop of the (virtual) 7th IWA/WEF Water Resource Recovery Modelling Seminar WRRmod2021. It is the result of a collective effort by industry practitioners and academics who are gratefully acknowledged for all their input. The work of Niels Nicolaï and Peter Vanrolleghem was supported by the Natural Sciences and Engineering Research Council of Canada Discovery Grant RGPIN-2021-04347 towards digital twin based control of water resource recovery facilities – Methods supporting the use of adaptive hybrid digital twins.

DECLARATION OF COMPETING INTEREST

The authors declare that they have no known competing financial interests or personal relationships that could influence the work reported in this paper.

DATA AVAILABILITY STATEMENT

All relevant data are included in the paper or its Supplementary Information.

REFERENCES

Afshari, H. H., Gadsden, S. A. & Habibi, S. 2017 Gaussian filters for parameter and state estimation: a general review of theory and recent trends. *Signal Processing* **135**, 218–238. doi:10.1016/j.sigpro.2017.01.001.

Akroyd, J., Mosbach, S., Bhave, A. & Kraft, M. 2021 Universal digital twin – A dynamic knowledge graph. *Data-Centric Engineering* **2** (4). doi:10.1017/dce.2021.10.

Alex, J., Hübner, C. & Förster, L. 2020 Planning, testing and commissioning of automation solutions for wastewater treatment plants using simulation. *IFAC-Papers OnLine* **53**, 16665–16670. doi:/10.1016/J.IFACOL.2020.12.1084.

AutomationML 2021 Available from: https://www.automationml.org/ (Accessed December 29, 2021).

Aymerich, I., Rieger, L., Sobhani, R., Rosso, D. & Corominas, L. 2015 The difference between energy consumption and energy cost: modelling energy tariff structures for water resource recovery facilities. *Water Research* **81**, 113–123. doi:10.1016/j.watres.2015.04.033.

Bartos, M. & Kerkez, B. 2021 Pipedream: an interactive digital twin model for natural and urban drainage networks. *Environmental Modelling and Software* **144**, 105–120. doi:10.1016/j.envsoft.2021.105120.

Bolton, A., Butler, L., Dabson, J., Enzer, M., Evans, M., Fenemore, T., Harradence, F., Keaney, E., Kemp, A., Luck, A., Pawsey, N., Saville, S., Schooling, J., Sharp, M., Smith, T., Tennison, J., Whyte, J., Wilson, A. & Makri, C. 2018 *The Gemini Principles: Guiding Values for the National Digital Twin and Information Management Framework*. Centre for Digital Built Britain and Digital Framework Task Group. doi:10.17863/CAM.32260.

Bonabeau, E. 2009 Decisions 2.0: the power of collective intelligence. *MIT Sloan Management Review* **50** (2), 45–52.

Brendelberger, M., Christmann, U., Dubovy, M., Gundlach, C. S., Hartmann, W., Hoelzke, U., Landa, D. V., Mertens, M., Modersohn, A., Nothdurft, L., Rahm, J., Schüller, A., Temmen, H. & Zgorzelski, P. 2019 *Digital Twin in Process Industry, NAMUR – Interessengemeinschaft Automatisierungstechnik der Prozessindustrie E.V.* Technical Report. Leverkusen, Germany, p. 4.

Bullard, M. G., Widdowson, M., Salazar-Benites, G., Heisig-Mitchell, J., Nelson, A. & Bott, C. 2019 Managed aquifer recharge: transport and attenuation in a coastal plain aquifer. In: *World Environmental and Water Resources Congress 2019: Groundwater, Sustainability, Hydro-Climate/Climate Change, and Environmental Engineering* (Bullard, M. G. ed.). American Society of Civil Engineers, Reston, VA, pp. 108–120.

Chinesta, F., Cueto, E., Abisset-Chavanne, E., Duval, J. L. & El Khaldi, F. 2020 Virtual, digital and hybrid Twins: a new paradigm in data-based engineering and engineered data. *Archives of Computational Methods in Engineering* **27**, 105–134. doi:10.1007/s11831-018-9301-4.

Corominas, L., Garrido-Baserba, M., Villez, K., Olsson, G., Cortés, U. & Poch, M. 2018 Transforming data into knowledge for improved wastewater treatment operation: a critical review of techniques. *Environmental Modelling & Software* **106**, 89–103. doi:10.1016/j.envsoft.2017.11.023.

Curl, J. M., Nading, T., Hegger, K., Barhoumi, A. & Smoczynski, M. 2019 Digital twins: the next generation of water treatment technology. *Journal American Water Works Association* **111**, 44–50. doi:10.1002/awwa.1413.

de Almeida Martins, J. P., Nilsson, M., Lampinen, B., Palombo, M., While, P. T., Westin, C.-F. & Szczepankiewicz, F. 2021 Neural networks for parameter estimation in microstructural MRI: application to a diffusion-relaxation model of white matter. *NeuroImage* **244**. doi:10.1016/j.neuroimage.2021.118601.

De Mulder, C., Flameling, T., Weijers, S., Amerlinck, Y. & Nopens, I. 2018 An open software package for data reconciliation and gap filling in preparation of water and resource recovery facility modeling. *Environmental Modelling & Software* **107**, 186–198. doi:10.1016/j.envsoft.2018.05.015.

DEXPI – Data Exchange in the Process Industry 2021 Available from: https://dexpi.org/ (accessed December 29, 2021).

Dickerson, S. T. & Butler, A. 2018 Resolve workforce challenges to ensure future success at water and wastewater utilities. *Opflow* **44**, 8–9. doi:10.1002/opfl.1063.

DTDL – Digital Twin Definition Language 2021 Available from: https://github.com/Azure/opendigitaltwins-smartcities (accessed December 29, 2021)

Eerikäinen, S., Haimi, H., Mikola, A. & Vahala, R. 2020 Data analytics in control and operation of municipal wastewater treatment plants: qualitative analysis of needs and barriers. *Water Science and Technology* **82** (12), 2681–2690. doi:10.2166/wst.2020.311.

Farazi, F., Salamanca, M., Mosbach, S., Akroyd, J., Eibeck, A., Aditya, L. K., Chadzynski, A., Pan, K., Zhou, X., Zhang, S., Lim, M. Q. & Kraft, M. 2020 Knowledge graph approach to combustion chemistry and interoperability. *ACS Omega* **5** (29), 18342–18348.

Fenu, A., Wambecq, T., de Gussem, K. & Weemaes, M. 2020 Nitrous oxide gas emissions estimated by liquid-phase measurements: robustness and financial opportunity in single and multi-point monitoring campaigns. *Environmental Science and Pollution Research* **27**, 890–898. doi:10.1007/s11356-019-07047-0.

FIWARE – The Open Source Platform for Our Smart Digital Future 2021 Available from: https://www.fiware.org/ (accessed December 29, 2021).

Fuller, A., Fan, Z., Day, C. & Barlow, C. 2020 Digital twin: enabling technologies, challenges and open research. *IEEE Access* **8**, 108952–108971. doi:10.1109/ACCESS.2020.2998358.

Glaessgen, E. & Stargel, D. 2012 The digital twin paradigm for future NASA and U.S. Air force vehicles. In *Proceedings of 53rd AIAA/ASME/ASCE/AHS/ASC Structures, Structural Dynamics and Materials Conference – Special Session on the Digital Twin.*

Grieves, M. 2014 *Digital Twin: Manufacturing Excellence Through Virtual Factory Replication.* NASA, Washington, DC, USA, White Paper 1.

GWRC 2021 *The Digital Water Utility of the Future: Whitepaper on the Enablers, Applications and Risks of Digitalisation for the Global Water Research Coalition.* Technical Report. Unley, Australia, p. 37.

Haimi, H., Mulas, M., Corona, F. & Vahala, R. 2013 Data-derived soft-sensors for biological wastewater treatment plants: an overview. *Environmental Modelling & Software* **47**, 88–107. doi:10.1016/j.envsoft.2013.05.009.

Heinonen, M., Jokelainen, M., Fred, T., Koistinen, J. & Hohti, H. 2013 Improved wet weather wastewater influent modelling at viikinmäki WWTP by on-line weather radar information. *Water Science and Technology* **68** (3), 499–505. doi:10.2166/wst.2013.213.

ISO – International Organisation for Standardization 2018 Organization and digitization of information about buildings and civil engineering works, including building information modelling (BIM) — Information management using building information modelling – Part 1: Concepts and principles – ISO 19650-1:2018. (accessed December 29, 2021).

IWA 2021a *Digital Water: Operational Digital Twins in the Urban Water Sector.* White Paper. London, UK, p. 17.

IWA 2021b *Digital Water: The Value of Meta-Data for Water Resource Recovery Facilities.* White Paper. London, UK, p. 15.

IWA & Xylem Inc 2019 *Digital Water Report.* Technical Report. London, UK, p. 44.

Jain, P., Poon, J., Singh, J. P., Spanos, C., Sanders, S. R. & Panda, S. K. 2020 A digital twin approach for fault diagnosis in distributed photovoltaic systems. *IEEE Transactions on Power Electronics* **35** (1), 940–956. doi: 10.1109/TPEL.2019.2911594.

Johnson, B. R., Kadiyala, R., Owens, G., Mak, Y. P., Grace, P., Newbery, C., Sing, S., Saxena, A. & Green, J. 2021 Water reuse and recovery facility connected digital twin case study: Singapore PUB's Changi WRP process, control, and hydraulics digital twin. In: *Proceedings of WEFTEC 2021*, Chicago, USA.

Jones, D., Snider, C., Nassehi, A., Yon, J. & Hicks, B. 2020 Characterising the digital twin: a systematic literature review. *CIRP Journal of Manufacturing Science and Technology* **29**, 36–52. doi:10.1016/j.cirpj.2020.02.002.

Kadiyala, R. & Macintosh, C. 2018 *Leveraging Other Industries – Big Data Management (Phase I).* Available from: https://www.waterrf.org/

Karmous-Edwards, G., Conejos, P., Mahinthakumar, K., Braman, S., Vicat-Blanc, P. & Barba, J. 2019 *Smart Water Report 2019: Foundations For Building A Digital Twin For Water Utilities.* Available from: www.swan-forum.com.

Le, Q. H., Verheijen, P. J. T., van Loosdrecht, M. C. M. & Volcke, E. I. P. 2018 Experimental design for evaluating WWTP data by linear mass balances. *Water Research* **142**, 415–425. doi:10.1016/j.watres.2018.05.026.

Lee, D. S., Vanrolleghem, P. A. & Park, J. M. 2005 Parallel hybrid modeling methods for a full-scale cokes wastewater treatment plant. *Journal of Biotechnology* **115**, 317–328. doi:10.1016/j.jbiotec.2004.09.001.

Le Moullec, Y., Potier, O., Gentric, C. & Leclerc, J. P. 2011 Activated sludge pilot plant: comparison between experimental and predicted concentration profiles using three different modelling approaches. *Water Research* **45**, 3085–3097. doi:10.1016/j.watres.2011.03.019.

Maktabifard, M., Zaborowska, E. & Makinia, J. 2019 Evaluating the effect of different operational strategies on the carbon footprint of wastewater treatment plants – case studies from northern Poland. *Water Science and Technology* **79** (11), 2211–2220. doi:10.2166/wst.2019.224.

Mamais, D., Noutsopoulos, C., Dimopoulou, A., Stasinakis, A. & Lekkas, T. 2015 Wastewater treatment process impact on energy savings and greenhouse gas emissions. *Water Science and Technology* **71** (2), 303–308. doi:10.2166/wst.2014.521.

Michael, J. & Wortmann, A. 2021 Towards development platforms for digital twins: A model-driven low-code approach. In: *Advances in Production Management Systems. Artificial Intelligence for Sustainable and Resilient Production Systems. APMS 2021. IFIP Advances in Information and Communication Technology*, Vol. 630 (Dolgui, A., Bernard, A., Lemoine, D., von Cieminski, G. & Romero, D. eds.). Springer, Cham. Available from: https://doi-org.acces.bibl.ulaval.ca/10.1007/978-3-030-85874-2_35

Muschalla, D. 2008 Optimization of integrated urban wastewater systems using multi-objective evolution strategies. *Urban Water Journal* **5** (1), 59–67. doi: 10.1080/15730620701726309.

Neto, O. B. L., Haddon, A., Aichouche, F., Harmand, J., Mulas, M. & Corona, F. 2021 Predictive control of activated sludge plants to supply nitrogen for optimal crop growth. *IFAC-PapersOnLine* **54** (3), 200–205. doi:10.1016/j.ifacol.2021.08.242.

Newhart, K. B., Holloway, R. W., Hering, A. S. & Cath, T. Y. 2019 Data-driven performance analyses of wastewater treatment plants: a review. *Water Research* **157**, 498–513. doi: 10.1016/j.watres.2019.03.030.

OGC – Open Geospatial Consortium 2021 Available from: http://www.opengeospatial.org/ (accessed December 29, 2021).

Olsson, G. & Newell, B. 1999 *Wastewater Treatment Systems.* IWA publishing, London, UK.

Olsson, G., Carlsson, B., Comas, J., Copp, J., Gernaey, K. V., Ingildsen, P., Jeppsson, U., Kim, C., Rieger, L., Rodríguez-Roda, I., Steyer, J.-P., Takács, I., Vanrolleghem, P. A., Vargas, A., Yuan, Z. & Åmand, L. 2014 Instrumentation, control and automation in wastewater – from London 1973 to narbonne 2013. *Water Science and Technology* **69** (7), 1373–1385. doi:10.2166/wst.2014.057.

Pantelides, C. & Renfro, J. 2013 The online use of first-principles models in process operations: review, current status and future needs. *Computers & Chemical Engineering* **51**, 136–148. doi:10.1016/j.compchemeng.2012.07.008.

Patwardhan, S. C., Prakash, J. & Shah, S. L. 2007 Soft sensing and state estimation: review and recent trends. *IFAC Proceedings Volumes* **40** (19), 65–72. doi: 10.3182/20071002-MX-4-3906.00012.

Pedersen, A. N., Borup, M., Brink-Kjær, A., Christiansen, L. E. & Mikkelsen, P. S. 2021 Living and prototyping digital twins for urban water systems: towards multi-purpose value creation using models and sensors. *Water* **13**, 592. doi:10.3390/W13050592.

PHES-ODM – The Public Health Environmental Surveillance Open Data Model 2021 Available from: https://github.com/Big-Life-Lab/PHES-ODM (accessed December 29, 2021)

Poch, M., Garrido-Baserba, M., Corominas, L., Perelló-Moragues, A., Monclús, H., Cermerón-Romero, M., Melitas, N., Jiang, S. C. & Rosso, D. 2020 When the fourth water and digital revolution encountered COVID-19. *Science of the Total Environment* **744**, 140980. doi:10.1016/j.scitotenv.2020.140980.

Quaghebeur, W., Torfs, E., Nopens, I. & De Baets, B. 2022 Hybrid differential equations: integrating mechanistic and data-driven techniques for modelling of water systems. *Water Research* **213**, 118166. doi.org/10.1016/j.watres.2022.118166.

Quay, A., Fiske, P. S. & Mauter, M. S. 2021 Recommendations for advancing FAIR and open data standards in the water treatment community. *ACS EST Engg.* doi:10.1021/acsestengg.1c00245.

Rahman, A., Belia, E., Kirim, G., Hasan, M., Borzooei, S., Santoro, D. & Johnson, B. 2021 Digital solutions for continued operation of WRRFs during pandemics and other interruptions. *Water Environment Research* **93** (11), 2527–2536. doi:10.1002/wer.1615.

Rieger, L., Gillot, S., Langergraber, G., Ohtsuki, T., Shaw, A., Takács, I. & Winkler, S. 2012 *Guidelines for Using Activated Sludge Models*. Scientific and Technical report. IWA Publishing, London, UK.

Ruohomäki, T., Airaksinen, E., Huuska, P., Kesäniemi, O., Martikka, M. & Suomisto, J. 2018 Smart city platform enabling digital twin. In: *Proceedings International Conference on Intelligent Systems (IS)*. IEEE, pp. 155–161. doi: 10.1109/IS.2018.8710517.

Russo, S., Lürig, M., Hao, W. J., Matthews, B. & Villez, K. 2020 Active learning for anomaly detection in environmental data. *Environmental Modelling & Software* **134**, 104869. doi:10.1016/j.envsoft.2020.104869.

Samad, T. & Mathur, A. 1992 Parameter estimation for process control with neural networks. *International Journal of Approximate Reasoning* **7** (3–4), 149–164. doi:10.1016/0888-613X(92)90008-N.

Spindler, A. & Vanrolleghem, P. A. 2012 Dynamic mass balancing for wastewater treatment data quality control using CUSUM charts. *Water Science and Technology* **65** (12), 2148–2153. doi:10.2166/wst.2012.125.

Stentoft, P. A., Munk-Nielsen, T., Vezzaro, L., Madsen, H., Mikkelsen, P. S. & Møller, J. K. 2018 Towards model predictive control: online predictions of ammonium and nitrate removal by using a stochastic ASM. *Water Science and Technology* **79** (1), 51–62. doi:10.2166/wst.2018.527.

Stentoft, P. A., Vezzaro, L., Mikkelsen, P. S., Grum, M., Munk-Nielsen, T., Tychsen, P., Madsen, H. & Halvgaard, R. 2020 Integrated model predictive control of water resource recovery facilities and sewer systems in a smart grid: example of full-scale implementation in Kolding. *Water Science and Technology* **81**, 1766–1777. doi:10.2166/WST.2020.266.

Stentoft, P. A., Munk-Nielsen, T., Møller, J. K., Madsen, H., Valverde-Pérez, B., Mikkelsen, P. S. & Vezzaro, L. 2021 Prioritize effluent quality, operational costs or global warming? – using predictive control of wastewater aeration for flexible management of objectives in WRRFs. *Water Research* **196**, 116960. doi:10.1016/j.watres.2021.116960.

Therrien, J. D., Nicolaï, N. & Vanrolleghem, P. A. 2020 A critical review of the data pipeline: how wastewater system operation flows from data to intelligence. *Water Science and Technology* **82**, 2613–2634. doi:10.2166/WST.2020.393/778299/WST2020393.

TWI2050 - The World in 2050 2019 *The Digital Revolution and Sustainable Development: Opportunities and Challenges. Report Prepared by the World in 2050 Initiative*. International Institute for Applied Systems Analysis (IIASA), Laxenburg, Austria. Available from: www.twi2050.org.

von Stosch, M., Oliveira, R., Peres, J. & Feyo de Azevedo, S. 2014 Hybrid semi-parametric modeling in process systems engineering: past, present and future. *Computers and Chemical Engineering* **60**, 86–101. doi:10.1016/j.compchemeng.2013.08.008.

Wiedau, M., von Wedel, L., Temmen, H., Welke, R. & Papakonstantinou, N. 2019 ENPRO data integration: extending DEXPI towards the asset lifecycle. *Chemie Ingenieur Technik* **91**, 240–255. doi:10.1002/cite.201800112.

Wilkinson, M., Dumontier, M., Aalbersberg, I., , Appleton, G., Axton, M., Baak, A., Niklas Blomberg, N., Boiten, J.-W., Bonino da Silva Santos, L. O., Bourne, P. E., Bouwman, J., Brookes, A. J., Clark, T., Crosas, M., Dillo, I., Dumon, O., Edmunds, S., Evelo, C. T., Finkers, F., Gonzalez-Beltran, A., Gray, A. J. G., Groth, P., Goble, C., Grethe, J. S., Heringa, J., 't Hoen, P. A. C., Hooft, R., Kuhn, T., Kok, R., Kok, J., Lusher, S. J., Martone, M. E., Mons, A., Packer, A. L., Persson, B., Rocca-Serra, P., Roos, M., van Schaik, R., Sansone, S.-A., Schultes, E., Sengstag, T., Slater, T., Strawn, G., Swertz, M. A., Thompson, M., van der Lei, J., van Mulligen, E., Velterop, J., Waagmeester, A., Wittenburg, P., Wolstencroft, K., Zhao, J. & Mons, B. 2016 The FAIR guiding principles for scientific data management and stewardship. *Scientific Data* **3**, 160018. doi:10.1038/sdata.2016.18.

Wright, L. & Davidson, S. 2020 How to tell the difference between a model and a digital twin. *Advanced Modeling and Simulation in Engineering Sciences* **7**, 1–13. doi:10.1186/S40323-020-00147-4.

First received 19 January 2022; accepted in revised form 18 March 2022. Available online 31 March 2022